ATHEISM
GENETICS TO GEOLOGY
AND MUCH MORE SCIENCE

ATHEISTIC
THOUGHT
THAT
GOES BEYOND
VERAGE
ONCEPT
IS REAL
WORLD

D1379821

MAURICE DE BONA, JR.

DESSERCO PUBLISHING

ACKNOWLEDGEMENTS

FOR INTERNET HELP
ARNE NEEMAN

FOR TYPESETTING
JULIE CREMEANS

ATHEISM
GENETICS TO GEOLOGY
AND MUCH MORE SCIENCE

by

MAURICE DE BONA, JR.

Copyright © 2011 by Maurice de Bona, Jr.

Library of Congress Control Number: 1-685104181
ISBN: 987-0-9852795-0-9 PAPERBOUND

Printed in the United States of America

First Edition: 2012

Desserco Publishing
PO Box 2433
Culver City, CA 90231

CONTENTS

ILLUSTRATIONS

1. INTRODUCTION

Happy, happy, happy. It is the firm belief of the author that the only purpose in life is to enjoy it. Now, there are many different forms of enjoyment, probably as many different forms of enjoyment as there are people. As we will see in the chapter on genetics, everyone has DNA that is slightly different; of course, that is except for identical twins.

Now, let us look at who might be interested in this book. I am told that approximately 40% of the people in the United States are creationists, but keep in mind that when people are polled on the subject of religion, they don't always tell the truth. Now, a creationist believes that God created the world and everything in it 6,000 years ago. But if not as few as 6,000 years, far less than the 5 million years ago when man evolved. For these people I do not believe that reading this book would be enjoyable.

Now we come to the next group of people, which I am also told represents approximately 40%. These people believe in gods, spirits and the hereafter. They may believe in evolution. It is hard not to believe in evolution, since it is exploding all around us today. Biologists have a hard time classifying all of the new species that have evolved. Also, modern science has developed quite accurate methods of dating rocks and fossils, which definitely proves evolution. These people might be interested in the book for the scientific information that it contains. This information is presented in a simple form that most everyone can understand.

Then there comes the last 20% who may or may not be atheists. Many of these people prefer to call themselves agnostics. Maybe it is because they really are not interested in going to church. Of course, there are closet atheists. They are those who go to church for business or social reasons. This is particularly true in smaller towns. So, that is one consensus of where the people of these United States of America stand religion-wise.

Now let us look at what the chapters of this book contain.

THE QUESTIONS

This chapter details a tragedy and presents a rational way of looking at it in relation to religion.

THE ORIGIN AND EVOLUTION OF GODS

This chapter summarizes James George Frazer's book "The Golden Bough", which is a compilation of accounts of primative religions as observed by early explorers.

BIBLE CONTRADICTIONS AND MYTHS

This chapter lists some of the best examples. It also tells why stories like "Noah's Ark" and "Jonah and the Whale" would not work.

THE LAW OF INERTIA AND HEAVEN

This chapter also talks about the Hubble space probe.

REALITY AND RELATIVITY IN LIFE

This chapter asks many questions about our existence. For example, the small atom and the large universe.

GEOLOGY AND EVOLUTION

Geologists have known about the periods of evolutionary emergence for quite some time, but it has only been recently that they discovered that radioactive isotopes found in rocks decay at a constant rate, and thereby they could date those rocks and the geologic periods. Also recent is their knowledge of continental drift.

GENETICS, DNA, RNA, ETC.

It has also been recently that scientists have unlocked the knowledge of what cells are made of and how they reproduce. This chapter summarizes this information in as simple a form as possible.

PLANT GENETICS

All living things are composed of cells. Bacteria are only single cells, but trees, for example, can contain many trillions of cells. This chapter tells how plant cells differ from animal cells.

SCIENCE, THE BRAIN AND HOW IT WORKS

First this chapter reviews some of the early ideas about the brain. Then we summarize the most recent ideas of how the brain works. But you must remember that we have a long way to go to understand memory and consciousness.

RELIGION AND THE CHURCH

This chapter looks at how religions are structured, and why religions are changing and why God still exists. Also, how churches regulate people.

MORALS IN THE WORLD TODAY

The author explores many of the morals in place today and gives his opinions regarding them.

IN CONCLUSION

This chapter summarizes previous chapters, presents additional information and draws additional conclusions. And lastly, the conclusion summarizes some of the other information presented in the book.

Day by day our unknown world is shrinking. We are living in a period where countries are moving closer together, socially and commercially. At the same time, each individual in this world has distinct differences and beliefs. With such a diversity of opinions, an increasing need presents itself for man to understand and respect the beliefs of others. It is with the hope of gaining a better understanding of why atheists think the way they do that this book was written.

During the Dark Ages, death was often accorded those not in agreement with the prevailing religious sect. Even now religious persecution is prevalent in many countries. Those of one faith find it difficult to understand the beliefs of other faiths. The people of this world need to make an added effort to comprehend the thinking of others. However, the contempt of past centuries is diminishing, and we are beginning to learn how to accept others for what they are, instead of what we want them to be.

Faith in a belief very often requires a great deal of courage. For many, faith has become not a possession, but a hope. While the greater part of the world holds firm to its inherent religions, faith in gods has weakened, and in some people it has disappeared, and they have become atheists.

This book discusses the reference points and fundamental background upon which non-belief is based. It attempts to show how and why atheistic thought differs from organized religion. It implies that atheistic thinking is real and that religious thought is in the realm of fairy tales. It does imply, however, that there is no absolute right or absolute wrong; that right and wrong are determined by the reference points upon which decisions are based. Conclusions, if they can be made, are left to the reader. The author definitely believes in freedom of religion, and if one is happy with one's religion, that is all that matters. Since atheistic thought differs from person to person, the views expressed here are not always consistent with those of other atheists.

There is little that is new in this writing. Until recently, the reader has had to wade through thousands of pages of literature to obtain the major points covered here. The book's highly condensed form has not, in many cases, permitted proper references. Comments are limited to the Jewish and Christian religions. One should keep in mind that close parallels often exist between the Judeo-Christian religions and other religions of the world. Also, exact figures are not important to the atheist.

This book is written in the masculine gender. When it says "man", "men" and "his", it also means "women" and "hers". Also, when the book reads "God" and "His", it is not intimating that God necessarily has a male or father image.

The author's belief is that reality, if there is such a thing, may be just beyond the comprehension of man, and consequently far different from what man perceives. The author attempts to carry his readers through a series of thought processes. These

4

processes will illustrate the possibility that "apparent reality" might only be a "fiction" produced by the human body's sensory organs. The author concludes that reality exists only in the eye, or sensory organs, of the observer, and nothing may be real, true or absolute.

Many factors contribute to the reasons why people have rejected religion and become atheists. This book does not come near to covering all of them. This is especially true when obscure reasons, such as rejection by the church, are considered. Whether or not Mary was a virgin is not important to this writing. Only the major factors leading to atheistic conversion are summarized. It may be asked, "What is an atheist, and why does he think the way he does?"

According to Webster's "New World College Dictionary", Second Edition, atheism is defined as follows: "Atheism is the belief that there is no God, or denial that God or gods exist." In other words, an atheist rejects religious beliefs and denies the existence of God. This does not mean that all atheists have a completely closed mind to the existence of some form of God. However, in light of their knowledge, the probability of a god existing seems doubtful to them. Consequently, the atheist chooses not to believe.

Webster also defines agnosticism as the "belief that the human mind cannot know whether there is a god or an ultimate cause, or anything beyond material phenomena." An agnostic is one who questions the existence of God, Heaven and any supernatural revelation. At the same time, he does not proclaim that he believes in them or does not believe in them. An agnostic draws no conclusions. This may be because the agnostic fears that if there is a god, and he rejects God, God will send him to Hell. The most likely reason, though, is because the agnostic feels that his knowledge is insufficient to come to a conclusion.

We do not perceive reality; our brains interpret it. Our vision appears to be fairly accurate. However, there are many aspects of vision that are not real. Our other senses are far less accurate. Scientific experiments have been conducted that show why the world that our brains reproduce in our heads is not entirely real. This book presents a few of those experiments.

A newborn baby looks out onto a nearly blank world. It takes weeks, months and years for its brain cells to be programmed and its memory processes developed to the point where it can perceive the whole picture.

Until recently, there has been very little research done on the functioning of the brain. Most of it was confined to observations after various parts of the brain were destroyed. It is only now that we are getting a clearer picture of its functions. This book explains those functions in a simplified form.

Even without any damage to the brain, the thoughts produced in it are often far from real. Unreal ideas come into our brains mostly by sight and sound. Once stored there, they are often interpreted as being real.

The human brain is very complex. It contains about 100 billion neurons, and more than nine times that amount of glial cells. The neurons receive small electrical charges from the sensory organs and send electrical charges to the muscles, etc. It is now believed that the glia cells store memory with calcium waves and send the commands to act to the neurons.

Brain cells cannot survive without a flow of blood to them. The blood carries oxygen and an array of different minerals and chemicals to nourish the cells. In other words, the brain is a material entity. It functions by material impulses coming to it and being processed in it, then signals are sent by it to control the material body.

Once our brain cells are programmed and memory processes are established, the brain can manufacture happenings inside itself that do not exist out beyond our body. An example would be UFO sightings. Also, most of our dreams are rather vague. However, some people have such vivid dreams that they believe their dreams are real.

Gods, spirits and the idea of a hereafter are defined as being nonmaterial. If they were material, then man's instruments could detect them. Consequently, as this book points out, the human concepts of gods, spirits and the hereafter are relegated to be simply ideas received by man's senses and stored by brain cells in his brain. Only the ideas exist, and not gods and spirits themselves.

To have thought, one must have a brain. Electrons, genes, cells and all other forms of matter that do not have a brain are unable to think, and, therefore, such matter has no real knowledge of its existence. Spirits, not being composed of matter, cannot have brains, and, therefore, cannot think or communicate.

Knowledge of how the human brain functions destroys the concept of gods, spirits and the hereafter, entering it other than as an idea. Again, gods, spirits and the hereafter are not composed of matter. The brain is a material entity. All inputs coming to it are composed of matter. The ideas of gods, spirits and the hereafter come into the brain by material sight and sound inputs.

There is a kind of environmentalist who hates to see the extinction of species. This may be because those people fear death and the preservation of species is another way of fighting it. Eventually, all species become extinct. Of course, it may take billions of years for that to happen. I was told that 550 million years ago there were only about 500 species. Today there are millions of species with maybe as much as 10 million species of insects alone. Along the way, many species evolved that were unsuited to the changing environment, and consequently they become extinct. There are at least six species of man who are no longer around. Someday, we (Homo sapiens) will be replaced by another species better adapted to the environment. The author believes that it is ridiculous for man to spend millions of dollars trying to save species from extinction. Unless the environment changes to favor such species, they will soon become extinct anyway. Again, nature has tried many combinations that are no longer around because they were not suited to the environment.

Man is probably the only animal that has any idea of gods and spirits. Anthropologists believe that man first learned to talk about 100,000 years ago. At this time, evolutionary changes occurred in man's throat, which enabled him to form vowels and other sounds. With the ability to form words, he was able to exchange complex ideas. However, these ideas had to be passed on from generation to generation through speech. It has only been a few thousand years since writing was developed and ideas could be put down in a more concrete form.

The last half-century has witnessed more advances in science than all of the years before them. Yet, the year 2000 may have only been the beginning of man's understanding of his universe. What the Bible and religion tell us is falling more and more into the realm of fairy tales. After all, bibles were conceived thousands of years ago when man was far less intelligent.

However, the majority of people today possess genetic neural programs in their brains to believe in religions. Still, at the same time, it is becoming harder and harder to accept those religions' beliefs.

2. THE QUESTIONS

There comes a time in life when questions are bought to mind about man's existence, and the existence of the god or gods (gods of Hinduism, for example) which he is told regulates the universe. For some, the questions come early in life. These people are often the type that fit their thoughts into a formula before coming up with an answer. The following formula might be one that they use.

If this is so, then that is so, or that is not so. This statement utilizes syllogistic reasoning or logic. It means that if one thing is held to be true, the other things that are compared to it are either true or false.

This kind of reasoning or logic fostered an examination of traditionally held religious ideas. It found discrepancies between what religion proclaimed to be and what religion was. For example, if God was strong, why did he appear weak? Certainly, he was unable to keep enemies of the Jewish people from murdering them, capturing them and holding them in slavery. If God was good, why did he cause so much bad to happen in the world? (The chapter entitled "Bible Contradictions" well illustrates this last point.)

Early atheists, such as Thomas Paine (1737-1809), Robert G. Ingersoll (1833-1899), and even as late as Bertrand Russell (1872-1970) had mainly this kind of reasoning to base their beliefs on. They lacked the scientific knowledge that we have today to prove the nonexistence of gods.

It was the many happenings in the world that these men observed that caused them to doubt the existence, or at least the benevolence, of God. The life experiences of man did not entirely add up to the truths of religious belief.

On the other hand, there are many people who go through life without thinking in detail about their existence, their religion, and their god or gods. It is only when something catastrophic happens to them or to others around them that they begin to examine the details of why. The following story, which appeared in the news media some years ago, illustrates such a catastrophe.

THE DAM

The time: October 1962. The place: Longarone, Italy, birthplace of my grandfather. Situated in the Italian Alps, the town, long noted as a tourist center, had increased in importance since completion of the Vaiont Dam in 1960. The 873-foot high structure, at the time the highest of its kind in the world, blocked the mouth of a canyon about two miles east of town. The Piave River meandered through town, southward towards Venice. Above rooftops rose the church steeples, ringing out their message, calling people to prayer.

The people of Longarone were a good people. They worked in the fields or ran small businesses in the town. Life centered around the churches. The people loved God or feared him, and for the most part lived by His commandments.

For weeks, heavy rains had fallen in the mountains causing minor landslides. People were becoming concerned. Thirty-five residents of the village of Pineda, just below the dam, were evacuated; 34 others refused to leave.

On the morning of October 9th, technicians at the dam decided to lower the reservoir's water level in anticipation of a larger slide. That night was a special night. Many had gathered in bars to see the 10:00 p.m. telecast of a championship soccer match between the Real Madrid team and the Glasgow Rangers.

Suddenly, at 10:43 p.m., an explosive roar thundered up and down the valley as the entire north face of Monte Toc collapsed into the reservoir behind the dam. In seconds, the lake, piling up in front of the slide, lapped up the villages on the opposite bank. Then it turned and roared over the dam, taking with it the superstructure, which included the control buildings and the model village where 43 technicians and their families lived.

In Longarone, the earth trembled and the windows shook. Those who got up to investigate felt a sudden, strong wind and heard a wild noise in the distance. Moments later, a 200-foot wall of water fell upon the city. The waters raced through town and up the hillside at its back. Receding back through town again, the waters turned downstream, destroying the smaller villages, such as Fae, that stood in its path.

In five minutes the flood vanished, leaving little except mud, rubble, and the dead. More than 2,000 men, women and children were annihilated. By one devastating blow of nature, or God, or of whatever power may be, a group of people living mostly by the law of the Lord had been destroyed. Public works minister Fiorentino Sullo called it a Biblical disaster like Pompeii. The flood tore bodies out of their graves. It left a plain of desolation almost unbelievable. Naked bodies hung from trees. Many more were buried wholly or partially in the mud or awash in streams. Many bodies were dismembered or unrecognizably crushed. Most of the corpses were calm and peaceful, appearing asleep.

Of the 2,900 people living in the devastated villages and surrounding countryside, only 700 survived. From the air, the valley looked like a major battlefield. Soldiers moved over the plain gathering the dead that were strewn everywhere. In the villages, wooden coffins were lined in rows. Priests, almost as numerous as the soldiers, moved among the coffins with their lips moving. From the air, planes sprayed disinfectants to prevent the spread of disease.

This was Longarone. This was Fae. This was a disaster of almost incomprehensible proportions. Why did it happen? Could it have been an error on the part of the engineers that designed the dam? They had been warned about the unstable condition of Monte Toc. On the other hand, was it an act of God? Was God responsible for the death of all those people? Is God responsible for all of the many disasters that have occurred before and after Longarone? Is God responsible for the recent tsunami?

A disaster, such as the Vaiont Dam flood, causes people to think in greater detail about their existence and the existence of God. They have been told that God is all-wise, all-powerful, all-loving and all-forgiving. Would a loving and forgiving god make his subjects suffer and die? Why should his children, who have worshipped him, feared him, and obeyed his commandments be subjected to this disaster? Is there a god at all? Couldn't the disaster have been an act of nature not caused by God? Couldn't this happening be the effect of the relationship and interaction of all the physical elements existing within that area?

The naturalists believe that the natural world, known and experienced scientifically, is all that exists and that there is no supernatural or spiritual creation, control or significance (*Webster's New World College Dictionary, Second Edition*).

God has been presented through some religions as having a paternal image. As stated in the Bible, that image is the same as man's. Certainly today many people do not believe in this kind of a god. God's image has become one whose form cannot be defined. God no longer has a gender. God is supposed to have respected man sufficiently to have created him absolutely free, to live and die by his own strength and weakness. As will be seen later, the author does not believe that there is free will in the universe. He believes that one's actions are governed by sensory inputs interacting with memory processes in the brain.

Man might ask, if there is a god, does God want to prevent evil and cannot, or can He prevent it and will not? Does not God appear either impotent or wicked? Is man's existence necessary for God's glory and for God's ego? Does He need someone to build up to great heights and then destroy? Or is the destruction of many merely due to the interaction of the elements of nature? Man is told to love God, for God loves him. Man is told that God is infinitely good. Yet, how can God love man and be infinitely good if He treats man cruelly? Is it not cruel for God to send those souls who have not obeyed all of God's commandments to Hell? Is there a Heaven and a Hell? Does man have a soul? Could this God be bloodthirsty, dishonorable, immoral and fickle? Does He enjoy torturing His subjects? Doesn't He have all of the qualities and faults of man himself? Isn't it possible that He is an invention of man and exists only in the mind of man?

These are some of the questions on which man might think. Later chapters will attempt to answer them. It should be kept in mind that religion is in the realm of philosophy, for which there is no real proof or answer.

Returning to the syllogism, if God is good (if this is so), then he has compassion for man (then that is so), or God is not cruel, jealous, immoral or dishonorable (or that is not so). From the questions asked previously, does this statement sound true? It does not sound true to the atheist.

3. THE ORIGIN AND EVOLUTION OF GODS

For thousands of years, and probably hundreds of thousands of years, man has prayed to gods. Man could not hear, see or converse with these gods, yet he believed that they were as real as man himself. As time went by, the concept of God and religion changed. Some religions progressed more than others, while a few changed hardly at all. However, none of the religions that exist today can truly be considered primitive. Even the most savage creeds have thousands of years of change behind them. On the other hand, just as some plants and animals have survived for millions of years virtually unchanged, there are some forms of religion today that have undergone very little change. Through observing the creeds of these religions, one can construct an evolution of religion. Early explorers wrote detailed accounts of the primitive religions that they observed. These have been summarized in a number of tomes, such as Sir James George Frazer's The Golden Bough, which is published by the Macmillan Co., London.

Anthropologists believe that it was about 100,000 years ago that man underwent a genetic change, whereby his larynx lowered in his throat, enabling him to speak, forming vowels and consonants. With this ability, words could be formed. With words, religious doctrines could be postulated. At first these doctrines were very simple. As time went on, they became more complex.

It is interesting to note that when parrots talk, one can see their larynx being lowered in their throat before they say, "Polly wants a cracker."

Some of the most primitive religions of recent times include those practiced by the Fuegans of South America, the Brazilian Indians of the Tapajos and Cupari Rivers, the Indians of the Gran-Chico (also in South America) and the Oukanyama Negroes of Africa. These people had no idea of a supreme being. They did not have the fear of punishment (as at the hands of God), or a dread of the natural phenomena surrounding them.

Even among primitive tribes, who have no religion or concept of God, a belief often exists that there is a life after death

13

and that man has something beyond the body, called a spirit, which forsakes the body at death. The moment of death is often described as the moment when the breath leaves the body. Possibly belief in religion grew out of fear. Fear was generated from man's terror of the mysterious natural agencies that threatened life.

There were three stages in the evolution of primitive religious belief. In the first stage, the dead were thought to still be alive, even though their bodies no longer exhibited motion. In the second stage, physical death is realized as a fact. At the same time, it is thought of as being only temporary, followed by a resurrection of the body. In the final or third stage, the soul, which is considered a separate entity from the body, is believed to leave and never to return to the body after death.

Gods as we know them are unknown in the first stage of primitive belief. It is the corpses and bones of dead men that are worshipped and revered. As an example, before going on an expedition, some Western African tribes scraped small portions of bone from the skulls of their ancestors, mixed them with water and drank it. This action made them brave and fearless like their ancestors. As society progressed, more and more examples of this practice, called "eating the god", can be found. Early burial practices also reflect the origin of gods. The body is often kept in a hut, a tree or on a platform where it is protected. Each day it is offered food and drink. Sometimes it is covered with a few inches of dirt until the body decays. Then, the head and sometimes bones were removed, cleaned and worn by the widow and relatives. Such practices are still followed in remote areas, such as New Guinea. Still, in other tribes, the relatives eat all or part of the flesh of the dead as an act of devotion and preserved the bones. These people believe that by wearing the bones of the relative, that relative will protect them. Through the practice of ancestor worship, one can see the beginning of the modern concept of God. For the main part, the Chinese people never went beyond this stage of ancestor worship.

The early American Indian believed his soul was the part of his body that traveled from him in his dreams. Maybe he would take a long trip with one of his relatives who was dead. They would travel through beautiful lands and kill many animals. When he awoke, there were no animals beside him, and his arrows, tomahawk and hunting knife were unused. He also was well rested from his sleep. He concluded that it was not he that went hunting, but his soul, and when he died, his soul would go to the happy hunting ground to stay. The Indian's soul had a bodily form, not like the abstract soul in which Christians believe.

The early savages had no problem in thinking positively about conflicting thoughts. They concentrated on only one thought at a time. With this type of thinking, religious beliefs did not have to add up to a truth.

The beginning of the second stage of religion coincided roughly with the beginning of the Neolithic Age, some 8,000 years B.C. This second stage was often accompanied by burial of the body. Burial reflected a certain fear of the dead. Man dreaded the return of the corpse. He also feared that the ghosts of the dead would haunt him in his dreams. He put heavy tombstones on the graves to keep the body from rising. Men buried their dead to get rid of them and to prevent the corpse from coming back and harming the survivors. Today, this idea has passed. The headstones that we use have no other significance than to mark the place where a person is buried.

In the past, when a deceased person was worshipped, it was not the corpse of that person that was worshipped. It was, instead, the person's gravestone. It is not always clear whether the belief was that the soul left the body and entered the stone or that the stone was the only visible reminder of the body. One way or another, these stones came to be recognized as gods. The stones became the objects of habitual worship. They were anointed with oil, milk and blood and given food and drink. Sometimes they were draped with cloth. As time went on, some stones were carved in the form of idols. As late as the nineteenth century, the use of sacred stones in Europe was not uncommon. One should keep in mind that Jahweh, the god of the Jews, was a sacred stone. Blood and other liquids poured over the ground at graves or on altar stones soon was absorbed by the ground or stones. This gave the impression that the ghost or god was sucking it in or drinking it.

In the beginning, blood sacrifices were made to the sacred stones. The Aztecs smeared warm blood from human victims on the idols' faces and threw pieces of palpitating human flesh into their mouths. As people became more civilized, they substituted animal sacrifices for human ones. Finally, such substitutes as red lead and cochineal or red ochre were smeared on the idols as a blood substitute. This practice is followed by the Hindus today.

When primitive man killed a beast, it was done at the time of a festival, and everyone joined in the feast. They all believed that they were eating in fellowship with their gods. Originally, humans were sacrificed and eaten, but again, as man became more civilized, he substituted sheep and oxen for the offerings.

The third stage in religious belief happened at about the time of the Bronze Age, approximately 3,000 years B.C. With it came the idea of the immortality of the soul. The body was of no further use after death. It was burned in the same manner as the Hindus do today. Burning was the ultimate effort in getting rid of the corpse. The ghost of the dead could still appear in dreams, but it was only the soul and not the dead person.

One may ask then, why are Christians still burying their dead? The answer is fairly easy. It is due to the idea that Christ would raise the bodies from the grave on Resurrection Day.

In India, bodies of the dead were not burned until the thirteenth or fourteenth century B.C. After this time, and up until the time of Christ, the ashes of the dead were buried in the ground. Then came the idea of throwing ashes in a sacred river.

In almost every case, gods came from the distant past. At their time of origin, they were nothing more than deified men. Some ghosts, or dead persons, became more important and powerful than others. These were often the ghosts of chieftains or witch doctors who possessed special powers to heal, increase fertility, or protect the people. They were ghosts who were worshipped the most and later developed into gods. As time went on, these gods were thought of more and more as spirits, and less and less as dead men, until finally the human form was lost. These gods often were identified with the sun, the moon, or other great powers of nature. The most important gods usually are those who are the oldest and least known. However, their importance depended on how well their priests promoted them and how many marvelous fables had been told about them. When it is possible to trace gods back in history, almost every one has had a human origin.

The first gods were cruel and matched the men who devised them. They were selfish and greedy for blood. Man attributed his miseries and misfortunes to the anger of these gods.

The king was also a god. The gods in the heavens were his ancestors. The king derived his power directly from these heavenly gods. When a king was sacrificed, he was sacrificed to the offended spirits of his forefathers. The son is sacrificed to the father, and the father sacrifices his son to make him a god. Where else have we heard this?

Many primitive people killed their king when he was young, or at least before he became old and weak. They regarded him as a god on Earth that possessed the powers of nature. If he were allowed to grow old and feeble, the rains would become few, the crops thin, and the rivers would run dry. There were other kings that held their throne for one year only, being sacrificed at the time of the spring equinox (Easter). These kings were gods of cultivation. Their longevity was very short. To improve this situation, the king would step down from his throne for a few days in spring, while a substitute king took over just in time for the sacrifice.

It was thought that a dying god, human or animal, could carry away the sins of the people. The dying god was bruised for the people's iniquities, wounded for the people's transgressions, and killed by the people's god, Jahweh, instead of them for the sins that they, the people, had committed. Without the shedding of blood, there is no remission of sin. The idea of sin, however, was a later concept. Early man thought little or nothing about sin. It was only after religion established an etiquette of the gods that it was possible to sin against them.

Agriculture was very important for the survival of people in ancient times. Then, as today, an important factor in raising crops was cultivation of the fields in which the crops were to be planted. Cultivation removes all other plants that would normally compete with the crop and hinder its growth. When primitive man buried his dead, he cultivated the land by digging the grave. He brought offerings of food and drink and made blood sacrifices at the grave. All these sacrifices fertilized the ground. Man soon noticed that the seeds, contained in the food that he brought as offerings, grew faster and larger than the plants in their natural environment. Primitive man thought that it was the spirit in the grave that was making these plants grow large and tall to repay him for his offerings. As a result of these observations, there developed throughout the world the practice of deliberately slaying a man at the time of spring planting to insure a vigorous and luxuriant crop.

To early man, Heaven, or the other world, was much more real than it is today. Man was quite willing to give up his life on earth by being sacrificed and then pass into Heaven. Now man is more skeptical. He is far less willing to exchange the comforts of this world for the uncertainty of the next.

The Konds of Orissa in East India had special victims called Meriahs who were often kept for years until needed for sacrificial purposes. Meriah men often had wives who were also Meriahs. If they had children, they were raised as Meriah victims. When a Meriah was killed at planting time, a delegate from each village carried a piece of the Meriah's flesh to the priest, who divided it in half. One piece was buried in the ground. The other was divided so each household had a small piece to bury in their field. Before this slaughter, the Meriah, like Christ, was anointed with oil.

The feast of the Mexican God, Tezcatlipoca, took place about the time of the Christian Easter. A year before the feast, a young man of unblemished body was chosen to be the god-king. He was given every luxury and treated as the deity himself. He was attended by eight royal pages, and everywhere he went, the people bowed down to him. Twenty days before the festival, four noble maidens bearing the names of goddesses were given to him as brides. The feast lasted five days, like many other sacrificial feasts around the world. During this time, the real king stayed in his palace. This clearly showed that the god-king was a substitute victim. The king's court attended to the victim's every desire. On the last day, he was ferried across the lake to a small pyramid temple. After ascending the last step, he was seized and held down to a sacrificial stone while the priest cut out his heart with a stone knife and offered it to the sun god. His head was hung among the skulls of previous victims, and his arms and legs were cooked for the lord's feast. The Portraj festival of southern India closely resembled the Mexican festival. This, again, suggests that the people of the Americas migrated across an Alaskan land bridge from Asia during a past glacial period. This festival is characteristic of god-eating ceremonies. On the first day of the five-day festival, a priest, or Portraj, who bears the same name as the cultivation god, was chosen. On the second day, a sacred buffalo was sacrificed by striking off its head with a single blow. The meat was then divided and buried by the cultivators in their fields. The blood and entrails were collected in a basket, over which the priest hewed a live kid. The basket was then placed on the head of a naked man, who ran around the boundaries of the village throwing the pieces right and left. The third and fourth days were witnessed by more sacrifices of buffalo and sheep. Also, on the fourth day, the women walked alone to the temple, dressed only in boughs of trees. The description goes on and on.

When the worshippers tore to bits a live sheep or goat, drank its blood and ate its flesh, they believed they were actually eating the god. By doing so, the god's properties would be imparted to them. The ancient Irish devoured the bodies of their dead fathers for the same reason. This was done to gain the courage and other qualities of the relative and to preserve the body and soul in a kindred body. Early man believed that the virtue of whatever he ate passed into his system. Therefore, he ate gods in the form of sacrificed human gods, animal substitutes, or bread and wine, which are symbols of flesh and blood. With this in mind, let us recall the Last Supper. Remember how the disciples ate the bread of Christ's flesh and drank the wine of his blood? Man believed that the vine represented the resurrection of God. The wine it produced was red like his blood, and when man drank it, he became inspired and intoxicated. He became "full of the god."

The common villagers were usually not allowed to present offerings to the chief ghosts or tribal gods. They had to be given to the chief, who would make the presentation, thus acting as a priest. Down through history, the priestly functions were often administered by one of the royal family. As time went on, the scriptures and religious ceremonies became more complicated, and the gap between the laity and the priesthood grew wider. Finally, the common man was doing little more than following religious customs that he did not really understand.

In Syria, there was a man-god named Adonis. He was originally selected and slain yearly. Later, he was slain and bewailed only in effigy. The mourning for Adonis, the fast, the sacrifice, the sacrament, and the atonement are almost identical to the Christian counterpart at Easter time.

A similar sacrifice was made by the Babylonians at the time of a festival called the Sacaea. At this time, a prisoner condemned to death ascended to the throne for five days, during which time he could eat, drink, order whatever he chose and even sleep with the king's concubines. At the end of this time, he was stripped of his robes, scourged by the people and crucified. Note the striking resemblance between this ceremony and the crucifixion of Christ. Remember that the people of the Holy Land were slaves in Babylonia for a time.

Paintings in the Egyptian tombs give a clear picture of how their religion evolved from ancient ancestor (or corpse) worship to final god worship. A whole pantheon of gods and goddesses evolved, with Osiris as the most important one. Osiris later became identified as an annually slain corn god. The Egyptian religion had a profound influence upon the Jewish people when they were held in Egypt as slaves. Among other things they borrowed from Egypt was the ark in which the sacred stone of the god, Jahweh, was kept. It was in front of this sacred stone that Saul became king of Israel. Jahweh rested in his ark at Gilgal among other gods or sacred stones of the Israelites.

The Jews were the first people to succeed in evolving a monotheistic religion. Up through the time of Hezekiah, Jahweh was only the highest of a group of Jewish gods. These gods were collectively called Baalim. Each seems to have consisted of a sacred stone standing under a sacred tree. Jahweh was said to be a jealous god, and in early times demanded human sacrifices. It is thought that the Molech, or king, who received human sacrifices was, in reality, god Jahweh. The lamb sacrificed at the time of Passover was a substitute for a human victim. It was the first-born child of a family that was sacrificed as a burnt offering at that time.

Jahweh was, by far, the most important Jewish god in the same way that Zeus was to Hellas and Osiris was to Egypt. The Jews carried him into battle to assure their victory. He was considered by many as their only power against destruction at the hands of larger nations. Even before the Baalim gods and their priests were destroyed, the practice of circumcision, which was substituted for human sacrifices after Jahweh became more civilized, must have assured Jahweh's power. According to legend, Jahweh wanted to kill Moses for not sacrificing his son. To save Moses, his wife took a knife, circumcised her son, and threw the bloody offering at Jahweh's feet to appease him.

Believing that an exclusive cult of pure Jahweh worship was all that would save Israel from her enemies, Josiah destroyed all that was Baal and all that was Asherah: temples, sacred stones and sacred trees. Then he murdered their priests, which numbered about 450. However, this could not save Israel. Three times in the following 30 years, the Babylonians captured and looted her towns, carrying off her people as slaves. The Babylonians destroyed Jahweh's temple and probably broke his sacred stone into many pieces.

One might think that Jahweh would die with the destruction of his temple and stone. Instead, it was only the beginning of a new, spiritual Jahweh, who was to become the nucleus of the largest monotheistic and cosmopolitan religion. This new Jahweh dwelled in the heavens. He was without image, and he was invisible to man. He became a sun god. Among other things, he was capable of sending lightning and thunder from the skies. One should remember that all great gods are associated with the sun. The Mikado of Japan was regarded as an incarnation of the sun. This transplantation of Jahweh to the heavens took place in the seventh century B.C.

Under monotheism, man shifted from a maternally structured society to a paternally structured society. God became selfish. Man, who God created, was God's property. God could do whatever he wanted to with man. He postulated principles for man to live by. He, himself, became bound to these principles in that God is truth, and God is justice. The image of God became separated from that of man. He was given an unpronounceable name, and to make his image was forbidden. He became the helping father who rescues man, punishes him for his disobedience, and is flattered by his praise.

The early Hebrews had little or no conception of a life beyond the grave, or a Heaven or Hell with rewards and punishments. There are no clear traces in books of the Old Testament, which were written before the Babylonian captivity, that described a doctrine of individual life beyond the grave. The Sabbath, legends of the creation, the fall of man, and the flood were all adapted by the Jews from the Babylonians during and after their captivity.

As time progressed, the idea began to evolve that, instead of making many human sacrifices, one man should die to carry away the sins and iniquities of the people for all times. This person was considered God in human form. It represented a sacrifice of himself unto himself.

There lived in Palestine from the second century B.C. until the second century A.D. a sect of people called the Essenes who followed the teachings of Buddha. The great similarities between Buddhism and the teachings of Christ indicate that his disciples must have been in contact with these people. It might be noted that the Buddhist religion reached its peak of influence about 200 years before Christ.

Like Christ, Buddha spent a period of time in the wilderness, where he was tempted, in vain, by the Devil. Like Christ's spectacular birth, Buddha was supposed to have been born of Maya, a king's daughter who became pregnant by a ray of sunshine. Buddha also came into the world with the purpose of destroying its misery.

At first it might seem strange that a human god such as Christ, without the power to keep mortals from killing him, should ever succeed to the position of a god. However, in light of all that has been said about sacrificial gods, it becomes evident that Christ was nothing more than one of them. He was to be the last god sacrificed, the last man to carry away the sins and iniquities of all the people for all time.

If there was a real person by the name of Jesus Christ, he certainly possessed the characteristics of a god sacrificed unto himself, who also combined the properties of a corn and wine god (cultivation god). During the Last Supper, Jesus said, "I am the true vine. Ye are the branches. I am the bread of life. Take, eat, this is my body. This is my blood of the New Testament." To this day, in the sacrament it is the custom of the church to eat the body of Christ in the form of bread and drink his blood in the form of wine. As previously mentioned, the Hebrews usually sacrificed a paschal lamb to Jahweh. Jesus is referred to in the Bible as "the lamb of God that taketh away the sins of the world."

The story of Christ is made of fragments taken from many god-slaying customs. He is the only begotten Son of God who died for the sins of mankind. Christ is treated as a god and temporary king. Like an Adonis, the people revere him. Branches are laid for him to walk upon. Herod dressed Christ in a gorgeous robe. He is brought before Pilate, who asks him, "Art thou the King of the Jews?" To this, Christ answers yes. After his death, Christ is mourned like an Adonis. Like the paschal lamb, Christ did not have his legs broken. The thieves crucified with him did. Instead, his side was pierced, whence flowed the blood of atonement in which the Christians were washed of sin. In actuality, blood does not flow after death. There is no heartbeat to pump it. As with corn and wine gods, Christ rose from the dead on the third day.

It is sure that Christ, if he did exist, was not the first of the saviors to die on the cross and carry with them out of this world the sins of the people for whom they were dying. Many times a god has been sacrificed himself, unto himself. Each time, the details of the sacrifice were similar, whether it was in ancient Egypt or ancient Mexico. Christ's birthday was fixed as the 25th of December. This is close to the shortest day of the year. It is also the birthday of Apollo, the sun god of Greek and Roman mythology.

Christianity did not develop all of a sudden with the death of Christ. The New Testament was written between 50 and 300 years after that time. It was written by men who were supposed to have lived with Christ and been his disciples. That was very unlikely, since in those days few people lived to be 50 years old. No wonder marked differences exist between their stories. As conceived, Christianity was to be a universal monotheistic religion acceptable to both Jews and Gentiles. It was free from the absurd legends of gods and goddesses, which were difficult for educated people to believe. Yet, it incorporated most of the beliefs of the existing religions. The trinity of the Christian pantheon, that is, the Father, the Son and the Holy Ghost, is reminiscent of the Egyptian love for triads or trinities. It included the Old Testament of the Jews. It included a sacrificial god, who would be acceptable to the Syrians and Babylonians. It contained "love thy neighbor" of the Buddhist Zoroastrians. It was as universal a religion as could be found in the Mediterranean.

At first, Christianity was the religion of slaves and poor people who lived in cosmopolitan seaports. It offered them some hope of salvation from their miserable existence. It was easy for Christianity to take over in the Mediterranean because, in actuality, it was little more than a compilation of the concepts already in existence. This new religion might have gone the way of other religions of its time, like Mithraism and Gnosticism, had not Constantine chosen it as the religion of the Roman Empire.

It was not long before the ruling classes saw the possibilities of using Christianity to control the masses. Under it, kings derived their authority directly from God, slavery was condoned, and the masses were compelled to obey the orders of the king.

Earlier, the author used the phrase, "if Christ did exist." Why should Christ not exist? The answer may lie in the amount of time, after the crucifixion, before the New Testament was written. Now, if one wanted to devise a perfect religion that all the world could believe in, it would be necessary to devise the perfect sacrificial god for it. Possibly, none of those gods sacrificed at the time were perfect enough to use. Also, it would be necessary to put the existence of this perfect corn and wine god back at least 50 years, so hardly anyone would be alive who knew whether or not he existed.

There is yet another reason to doubt the existence of Christ. The Romans were meticulous keepers of records. It has been said that there is no entry in the Roman journals to note the crucifixion of Christ. Of course, it could have happened and never been recorded.

At this point I want to mention that I only talked with Madeline Murray O'Hare once on the telephone, so I have no personal feelings about her. But from what I have heard, she truly deserved the title of "The Most Hated Woman in America." I will give her credit, though, for the extensive amount of work that she did researching records from the time of Jesus Christ. She only found one record of someone writing that he had heard of him. That is why I truly believe that Christ was only a story.

People today have little or no knowledge of the origin of gods. Consequently, its establishment as a basis for non-belief is almost nonexistent. However, if one concludes that God is man-made by our misguided ancestors, then in reality, there is no God at all.

4. BIBLE CONTRADICTIONS AND MYTHS

The Bible, the world's bestseller, was selected as the topic for this chapter because of its widespread use. Other religious writings, such as the Koran, could have been used equally well to illustrate discrepancies that result from comparing it to modern thinking. Bible contradictions were an early reason for non-belief in God. Later, scientific discoveries and the Theory of Evolution added kindling to the already growing doubts the Bible created. Early atheists, such as Thomas Paine and Robert G. Ingersoll, lacked much of the scientific evidence that we have today. In fact, it has been the last half century when most of these discoveries were made.

To find that the Bible contains many contradictions and errors is not surprising when it is realized how much information has been incorporated into the Bible through the centuries. The scriptures have sustained numerous changes, as is witnessed by many erasures on the manuscripts. The Old Testament was printed for the first time in 1488. At the present time, it is still being revised by various religions.

Many interpretations, often called concordances or commentaries, have been written to justify the Bible's apparent errors. The atheist is more inclined to accept what is written for face value only. He makes little or no attempt to justify the Bible passages by looking for hidden meanings.

Probably more than any other single reason why modern man has lost faith is because the teachings of the Bible and other holy books did not agree with what he found as truths in the world around him.

Of course, these great religious books were written thousands of years ago. They were handed down from generation to generation. During this time, man's knowledge and beliefs progressed and changed. Today, man is often faced with the problem of reconciling the discrepancies between these holy books and his present knowledge.

At one time, the Bible was believed to be the actual word of God, revealed by God to Moses and the prophets. Most religious people at present believe that it was written, instead, by numerous men who were inspired by God to write what they interpreted, in the light of their knowledge, to be the word of God. Because of its numerous apparent fallacies, one is told that the Bible should be read only in spirit and not in a literal sense.

Before determining what is and what is not in error, some assumptions must be made. God is assumed to be good and not evil, right and not wrong, and perfect, not imperfect. God has control, at least to some degree, over everything in the universe. It is assumed that God made the world, and he made man in his image to live upon it.

It is generally conceded by most Biblical authorities today that the Pentateuch, which contains the Ten Commandments, was not written by Moses. It was obviously authored by many people over a period of centuries. Most of the laws of the Pentateuch, or first five books of the Bible, were in general use hundreds of years before the time of Moses. This fact makes the "direct revelation from God" theory not plausible.

According to the Bible, the Earth and all living things on it were created by God in the year 4004 B.C. This date was determined by Archbishop Ussher from the Genesis narrative. Other chronologies place the date of creation up to 20,000 B.C. Hardly anyone today would believe that the world and man have been around for such a short time.

> *And God said, "Let there be a firmament in the midst of the waters, and let it separate the waters from the waters." And God made the firmament, and separated the waters which were under the firmament from the waters which were above the firmament: and it was so. And God called the firmament Heaven…* (Genesis I:6-8)

According to the Bible, God was supposed to have created the universe and everything in it out of nothingness in six days. He made the earth flat with a solid blue dome over it. He made the sun to revolve so it would give light by day, and the moon to revolve so it would give light by night. On the seventh day he rested. This story must be considered in the realm of fairy tales today.

God created man in his own image by blowing breath into dust, the Bible says. To early man, the essence of life was breath. He could make statues of clay that resembled himself, but they had no breath. It was only natural for man to believe that god blew breath into clay statues to make live animals.

From one of Adam's ribs, God fashioned a woman, whom he called Eve. God made Adam and Eve perfect, and at this time, there was no sin in the world. But the Devil, in the form of a serpent, tempted Eve, who in turn tempted Adam to eat an apple from the forbidden tree. In doing so, they sinned in the eyes of the Lord. As punishment for this sin, death was accorded to all men.

God created all living creatures and also the seed bearing plants. Also, it must have been quite a task to throw up all those high mountains in the world.

Before the middle of the seventeenth century, it was commonly thought that God created mankind and the higher organisms, but insects, frogs and the lower forms of life arose spontaneously from mud or decaying matter.

[SEE A BIBLE HISTORY ILLUSTRATION]

Contrary to the Biblical creation, geologists and anthropologists have found remains of man that date back several million years. Evidence also indicates that man evolved from earlier tree-dwelling animals and still much earlier from fishes. Man has undergone a gradual evolution to reach his present development and intelligence.

Methuselah is said to have lived to be very old. This is because at the time the world began, there wasn't much sin in it.

And all the days of Methuselah were nine hundred sixty and nine years; and he died. (Genesis V:27)

In actuality, man probably never lived longer than he does today. Very few people reached the age of 50 at the time of Christ.

A BIBLE HISTORY

4004 BC	THE CREATION [According to Ussher] ADAM [6th Day] EVE [Man Sinned]	925 BC · ISRAEL [Northern Kingdom] · JUDAH [Southern Kingdom]
	ABLE · CAIN	JEROBOAM [First of 19 Kings-About 200 Years] · REHOBOAM [First of 20 Kings-About 335 Years]
3770 BC	SETH ENOS	
	CAINAN	721 BC · ASSYRIAN CAPTIVITY
	MAHALALEEL	586 BC · BABYLONIAN CAPTIVITY
	JARED ENOCH	RETURN TO JERUSALEM
3320 BC	METHUSELAH	539 BC · END OF BABYLONIAN RULE
	LAMECH	
2515 BC	[The Deluge] NOAH	536 BC · ZERUBBABEL
		458 BC · EZRA
	SHEM · HAM · JEPHETH	445 BC · NEHEMIAH
1850 BC	ABRAHAM · Immigration to Africa & Arabia · Immigration to Asia Minor & Europe	HAGGAI ZACHARIAH
	ISAAC	MALACHI
	JACOB, ISRAEL & 12 Sons	333 BC · PERSIAN PERIOD · Oppression, Invasion, & Bloodshed
		167 BC · GREEK PERIOD
1650 BC	[Went into Egypt]	· JEWISH FREEDOM
1290 BC	MOSES [The Exodus] [40 Years Wandering]	63 BC · ROMAN RULE
	JOSHUA	6 BC · BIRTH of JESUS
	THE JUDGES	25 AD · BAPTISM of JESUS
		29 AD · CRUCIFICTION of JESUS
1020 BC	SAUL [First King of Israel]	50 AD / 300 AD · NEW TESTAMENT WRITTEN
1000 BC	DAVID	325 AD · CONSTANTINE Adopted Christianity as Roman State Religion
961 BC	SOLOMON	

Dates of events are approximate. They are interpreted from the Bible, and vary from one chronology to another.

Man is told that God, creator of the world and the universe, is supposed to be all-wise, all-powerful, all-loving, and he is revealed to us in the scriptures. However, the scriptures do not always reveal this kind of god. They often reveal one who is jealous, fickle, immoral, dishonorable, barbarous and cruel. In other words, he possesses all of the characteristics of man himself.

If there is a God, there is no reason why he should be perfect, and if the Bible was conceived through divine inspiration, there is no reason why it should be perfect either. However, it appears illogical to a nonbeliever that a work inspired by a great power like God could not be more perfect. After observing the numerous contradictions of the Bible, the nonbeliever is more inclined to think that it was written by ordinary men without divine guidance at a time when knowledge of the world was considerably less than it is today.

CONFLICTING BIBLE PASSAGES

1. Man was created before the animals; man was created after the animals.

> Then the Lord said, "It is not good that man should be alone; I will make him a helper fit for him."... (Genesis II:18) And out of the ground the Lord God formed every beast of the field, and every bird of the air, and brought them unto Adam to see what he would call them... (Genesis II:19)

> And God made the beasts of the earth according to their kinds, and the cattle according to their kinds... (Genesis I:25) And God said, "Let us make man..." (Genesis I:26) So God created man in his own image... (Genesis I:27)

2. God is all-powerful; God is not all-powerful.

> "...with God all things are possible." (Matthew XIX:26)

> And the Lord was with Judah, and he drove out the inhabitants of the mountain, but he could not

drive out the inhabitants of the plain, because they had chariots of iron. (Judges I:19)

3. God is loving; God is cruel.

...God is love... (1 John IV:16) The Lord is good to all; and his tender mercies are over all his works. (Psalms CXLV:9)

"...I will not pity, nor spare, nor have mercy, but destroy them." (Jeremiah XIII:14)

4. God is peaceful; God is warlike.

For God is not the author of confusion, but of peace... (1 Corinthians XIV:33)

The Lord is a man of war... (Exodus XV:3) Blessed be the Lord my strength, which teacheth my hands to war, and my fingers to fight. (Psalms CXLIV:1)

5. God says, "Thou shalt not kill." God says, "Thou shall kill".

Thou shalt not kill. (Exodus XX:13)

..."Thus said the Lord God of Israel, 'Put every man his sword by his side, and go to and fro from gate to gate throughout the camp, and slay every man his brother, and every man his companion, and every man his neighbor.'" (Exodus XXXII:27)

6. God forbids human sacrifices; God commands human sacrifices.

Take heed that you be not ensnared to follow them [the gentiles]... (Deuteronomy XII:30) ...for even their sons and their daughters have they burnt in the fire to their gods. (Deuteronomy XII:31)

He [God] said, "Take now thy son, thine only son Isaac, whom thou lovest, and get thee into the land of Moriah, and offer him there for a burnt offering upon one of the mountains of which I shall tell you." (Genesis XXII:2)

According to the scriptures, God spared Isaac's life at the last minute. Instead, a ram found trapped in a thicket was substituted for the burnt offering. The Bible contends that this episode was God's test of Abraham's faithfulness. Does God have so little knowledge of the feelings of man, whom he created, that he needs to run tests to know if man is faithful? This kind of test is the kind of test a man would make, not God.

7. Adultery is forbidden; adultery is sanctioned.

   Thou shalt not commit adultery. (Exodus XX:14)

   And the Lord said to Hosea, "Go again, love a woman who is beloved of a paramour and is an adulteress..." (Hosea III:1)

8. Divorce is restricted; divorce is permitted.

   But I say unto you, that whosoever shall divorce his wife, saving for the cause of fornication, causeth her to commit adultery; and whosoever marries a divorced woman commits adultery. (Matthew V:32)

   When a man has taken a wife, and married her, and it comes to pass that she finds no favor in his eyes... then let him write her a bill of divorcement, and give it into her hand, and send her out of his house. (Deuteronomy XXIV:1)

9. Intoxicating beverages are disapproved; intoxicating beverages are recommended.

   Wine is a mocker, strong drink is a brawler; and whosoever is deceived thereby is not wise. (Proverbs XX:1)

   "...wine which cheereth God and man...." (Judges IX:13) *Give strong drink unto him that is ready to perish, and wine to those that be of heavy heart.* (Proverbs XXXI:6) *Let them drink and forget their poverty, and remember their misery no more.* (Proverbs XXXI:7)

10. All men sin; those born of God do not sin.

   ...for there is no man that sinneth not... (1 Kings VIII:46)

   Whosoever is born of God doth not commit sin... (1 John III:9)

11. The godly shall not suffer; the godly will suffer.

   No evil shall befall the righteous... (Proverbs XII:21)

   Yea, and all that will live godly in Christ Jesus shall suffer persecution. (2 Timothy III:12)

12. Poverty is a blessing; riches are a blessing.

   ..."Blessed be the poor..." (Luke VI:20) *"...woe unto you that are rich..."* (Luke VI:24)

   The rich man's wealth is his strong tower, but the destruction of the poor is their poverty. (Proverbs X:15)

13. The dead shall be resurrected; the dead shall not be resurrected.

   ...the trumpet shall sound, and the dead shall be raised... (1 Corinthians XV:52)

   As the cloud is consumed and vanisheth away, so he that goeth down to the grave shall come up no more. (Job VII:9)

14. Jesus wants his disciples to hate; Jesus says that one who hates is a murderer.

   "If any man comes to me, and hates not his father, and mother, and wife, and children, and brethren, and sisters, yea, and his own life also, he cannot be a disciple." (Luke XIV:26)

   Whosoever hateth his brother is a murderer; and ye know that no murderer hath eternal life abiding in him. (1 John III:15)

15. Christ came to bring peace; Christ came to bring a sword.

And thou child, shall be called the prophet of the highest... (Luke I:76) *...to guide our feet into the way of peace."* (Luke I:79)

"Think not that I am come to bring peace on earth; I came not to bring peace, but a sword." (Matthew X:34)

Some of these conflicting passages have compared one passage from the Old Testament with another from the New Testament. Many feel that these two parts of the Bible represent different religions. It is obvious that the religious thinking of the Christian era differs from that which went before it. However, when I was a child, the Old Testament and the New Testament were one Bible. How things have changed.

"Think not that I am come to destroy the law [of God] [the Pentateuch] and the prophets; I am not come to destroy, but to fulfill. (Matthew V:17)

In so saying, Christ accepts the Jewish god for the Christian religion. At the time that the Old Testament was first conceived, the people believed that there was more than one god. However, since then, the modern Jewish and Christian religions have become monotheistic religions. That is, they believe in the doctrine that there is only one god. If this is so, there cannot be different gods for each religion, even if each religion interprets God differently. If there is only one god, and God is not born and does not die, then this one god must be the same one that existed during both the pre-Christian and Christian eras. If God is capable of revealing the truth to man, he should have been able to reveal the truth to pre-Christians as well as to Christians. In making these assumptions, one presupposes that there must be an absolute truth. If, in reality, there is no true or false and right or wrong, then God, if he does exist, would not be obliged to reveal information that was truthful or righteous.

One is told that God represents righteousness. Yet, He causes men to kill their neighbors. He causes the peasantry to be oppressed by their lords, who, in turn, are supposed to derive their powers from God. The aim of God, one is told, is to make men happy. Yet, man suffers much, enjoys little and dies. We are told that it is our sins that make God punish us.

Is God peace-loving? Is a god who commands his people to go out and kill his neighbors, every man, woman, infant and suckling (1 Samuel XV:3), peace-loving? Isn't God intolerable? He got angry because the Jews wanted something else to eat besides manna, which is similar to coriander seeds (Numbers XI:4-6)? Can God have ordered a man to throw a spear through the body of a woman to stop a plague (Numbers XXV:8)? Such happenings seem more like superstitions contrived in the mind of primitive man.

According to the Bible (Deuteronomy XIII:6-10), the god of the Jews ordered them to kill their wives, children and brothers if they worshipped any other gods. One can only conclude that this god was a bloodthirsty one. He took pleasure in giving no mercy. He ordered mothers to be ripped open with a sword and their babies butchered in their arms. This is not the god that man knows today. Has God changed, or have the men that created God changed?

The Bible says:

Thou shalt not suffer a witch to live. (Exodus XXII:18)

As a consequence of this Biblical reference, through the centuries, hundreds of thousands of innocent women were put to death. They were unfortunate enough to be marked with the imaginary crime of being a witch. The Bible must have ranked high during these times as the instrument for initiating brutality among people.

In the pre-Christian era, human sacrifices to the gods were common. The Bible documented these well.

Then the sprit of the Lord came upon Jephthah... (Judges XI:29) *And Jephthah vowed a vow unto the Lord, and said, "If thou shalt without fail deliver the children of Ammon into mine hands, then it shall be, that whatsoever cometh forth of the doors of my house to meet me, when I return in peace from the children of Ammon, it shall surely be the Lord's, and I will offer it up for a burnt offering."* (Judges XI:30-31) *And Jephthah came to Mizpah unto his house, and behold his daughter came out to meet him with timbrels and with dances; and she was his only child...* (Judges XI:34) *And it came to pass at the end of two months that she had returned unto her father, who did with her according to his vow, which he had vowed...* (Judges XI:39)

And Aaron shall lay both hands upon the head of the live goat, and confess over him all the inequities of the children of Israel, and all their transgressions, in all their sins, putting them upon the head of the goat, and shall send him away by the hand of a fit man into the wilderness; and the goat shall bear upon him all their inequities unto a land not inhabited... (Leviticus XVI:21-22)

Is not the sacrificing of livestock like offering a bribe in exchange for forgiveness of one's transgressions against God? Was it God who ordered the killing of one's brother, daughter or son? Or were these happenings the result of primitive people incorrectly believing that they were carrying out the orders of God?

When Moses fled from Egypt, the Midianites, who were closely related to the Israelites, gave him shelter (Exodus II:15). The Bible (Numbers XXXI:1-18) tells us how Moses repays their hospitality.

And the Lord spake unto Moses, saying, "Avenge the children of Israel of the Midianites..." ...and they slew all the males. ...and the children of Israel took all the women of Midian captive, and their little ones; and took the spoil of all their cattle, and all their flocks, and all their goods. And they burnt all their cities wherein they dwelt, and all their goodly castles, with fire. They took all of the spoil, and all of the prey, both of men and of beasts. And Moses was wroth... And Moses said unto them, "Have ye saved all the women alive? Now therefore kill every male among the little ones, and kill every woman that hath known man by lying with him. But all of the women children, that have not known a man by lying with him, keep alive for yourselves."

Is it possible that God could have given man the Ten Commandments and then ordered him to break most of them? How about the one, "Thou shalt not commit adultery?" What did David do about it?

> *And it came to pass in an eveningtide, that David arose from off his bed, and walked upon the roof of the king's house; and from the roof he saw a woman washing herself, and this woman was very beautiful to look upon.* (2 Samuel XI:2) *And David sent messengers, and took her; and she came in unto him, and he lay with her; for she was purified from her uncleanness; and she returned unto her house.* (2 Samuel XI:4) *And it came to pass in the morning that David wrote a letter to Joab, and sent it by the hand of Uriah [this woman's husband]. And he wrote in the letter, saying, "Set ye Uriah in the forefront of the hottest battle, and retire ye from him that he may be smitten, and die."* (2 Samuel XI:14-15)

After David committed adultery with this woman, he arranged for her husband to be killed so he could marry her. This, of course, was not a sin because David was the seed of God.

THE MYTHS

Of all the writings that fall into the category of Bible myths, "Noah and the Ark" must take the cake. Originally borrowed from the Babylonians, the story was adapted to the Old Testament.

God commands Noah to build an ark to house all the species of the world during the great flood that God was sending to rid the world of human wickedness. Noah was to take aboard a male and female of each species, plus enough food for all. Now this ark was measured in cubits. In feet it would be about 75 feet wide by 450 feet long by 45 feet high. At the time, there were thousands and thousands of species in the world. There were maybe more than 12,000 species of birds, 6,000 species of reptiles and 1,600 species of mammalia. This was not to mention one million other kinds of creatures such as insects. One would have to be very dumb to believe that all these species could fit on that little ark. Now this flood was supposed to have lasted for more than ten months. There is no way that this little ark could carry enough food for ten months. Also, many foods would spoil. Some species require special foods, which Noah could not obtain, and when the food ran out, the carnivores would eat the herbivores. It would be complete chaos.

And the waters prevailed exceedingly upon the earth, and all the high mountains that were under the whole heaven were covered. (Genesis VII:19)

Where all of this water came from to cover mountains 29,000 feet high is hard to imagine. Even if the flood had been local in nature, it would have left a record in the sedimentary rocks deposited at that time. As far as is known, geologists have not found evidence of such a deluge. There is also the vegetable kingdom. How many plants would have survived being covered with water for ten months? However, God also instructed Noah to collect the seeds from the seed bearing plants. That would have been a super task, if not impossible.

If almost all of life on Earth had been destroyed by a great food less than 6,000 years ago when Noah lived, how do you account for the fact that there is so much life on planet Earth today? It took billions of years for that life to evolve to where it is today.

We could go on and on with the impossibilities that Noah would have encountered, but that is enough.

Then there was "Jonah and the Whale". No one has to tell us that "Jonah and the Whale" would not work. After being swallowed by a whale, or a big fish, as the Bible says, he would have died. It would be impossible for him to survive for three days and three nights in the belly of a whale.

It is doubtful whether Jesus preached a sermon on the mount. John said that the sermon was on the plain. One should keep in mind that the mount is part of the myth of the sun god on his hill. The twelve apostles correspond to the Signs of the Zodiac.

> *Christ sat upon the Mount of Olives, and his disciples came unto him. And they asked him when he shall come, and when shall be the end of the world; and Christ replied, "Verily I say unto you, this generation shall not pass, till all these things take place."* (Mark XIII:3, 4, 30)

Many generations have passed, and the world is still here. Furthermore, if man doesn't blow it up, it will probably be here for many more billions of years.

There are many different Bibles. Each religion's Bible varies form the others. Most of the passages presented in this chapter are from the King James Version. They have been checked against the same passages in other Bibles. Often the wordings were found to vary, but the meanings appeared to remain the same. On the other hand, interpretations of Bible passages can vary greatly from one religion to another.

Those who do not analyze what the Bible is saying have no problem in believing it. They read each section by itself and for itself and make no comparisons. Those who can read hidden meanings from the scriptures can also justify discrepancies that may exist. The atheist finds it difficult to make these justifications. Many have left the ranks of the religious, solely because of the discrepancies they have found in the Bible.

5. THE LAW OF INERTIA AND HEAVEN

The law of inertia ranks high in importance when used to justify the beliefs of the atheist. The law of inertia is the prime factor governing the continuance of all functions of life. It explains in a natural way why matter tends to continue doing in the future what it is doing at present. It explains why man wants to continue his life in another everlasting world.

Webster's New College World Dictionary, Second Edition, defines inertia as "the tendency of matter to remain at rest if at rest, or if moving, to keep moving in the same direction, unless affected by some outside force."

This definition is not entirely true when considered in terms of nuclear building blocks. Electrons, protons, etc. are always in motion and never at rest. Consequently, these nuclear building blocks tend to travel in the future the same way as they are traveling at present. It is not possible for them to do so because other nuclear building blocks are ever present to exert opposing forces, thereby changing the direction of travel. For this reason, objects tend to take the path of least resistance. A body passing from one place to another or from one state to another chooses the route of least resistance.

Just as the law of inertia affects matter, it also affects everything composed of matter. Plants and animals want to continue living in the future because they are living at present. It takes a force to cause death. It takes a force to interrupt the continuing processes of cell reproduction.

Of all of the forces that hold man to religion, this desire for continuation after death is probably the greatest. The idea of eternal life in Heaven represents the ultimate continuation that the law of inertia could achieve. The principles of inertia, or continuance, apply to all the facets of life. Continuance stabilizes man's life. This desire for continuance has enabled cultures to persist as they have in the past, from generation to generation, for thousands of years. By observing these cultures in recent times, modern man has been able to establish an idea of how the concept of gods came to be.

To the atheist, it becomes evident that religion, as it is known today, was born out of man's desire for continuance. Contrary to this desire, matter appears to be constantly changing. The composition of living matter changes from moment to moment. Death of living matter is the result of an interruption of vital body functions necessary to sustain the functioning of the whole. The matter comprising the body can neither be created nor destroyed. It will continue on to eternity, changing from moment to moment.

The atheist accepts a universe without continuation because for him to believe in a hereafter is contrary to his knowledge of matter. To believe in a surviving soul not composed of matter is foreign to his experience. Everything that acts upon his senses is composed of matter, and a soul is nonmaterial. To the atheist, the soul is nothing more than the functioning of the whole body. When the body dies, the soul ceases to exist.

As previously stated, one of the laws of science is that matter can neither be created nor destroyed. That tells one that the universe has always been and will always be. The matter that is in our bodies now will continue in one form or another for eternity.

Also keep in mind that all of the matter that was in our bodies as a young person has been replaced by other matter by the time that we get old.

Early man had a much different idea of what heaven was. He didn't realize that there was a vast universe out there. We can now see the universe in much greater detail with such tools as the Hubble space telescope. We wonder if there is life out there, but life cannot travel those vast distances. Maybe some day robots can do it.

6. REALITY AND RELATIVITY IN LIFE

There was a time when religious men feared science. They believed that science would destroy the peace of mind that religion had given them. As the years passed, it became evident that science could not answer all of man's questions, and it still can't. Each individual is forced to evaluate the information available and choose a belief in or against the concept of God. Most people don't bother. God sounds good to them, so they don't evaluate science. The atheist chooses to believe in those scientific findings that point to the nonexistence of God, even though in doing so he has lost the promise of salvation after death. This was because scientific knowledge appeared to add up to a truth in his mind.

Scientific findings brought many questions to man. The solid world of early times began to disintegrate. Man began to wonder if it were truly possible to understand the world around him. Was there anything in it that was real? Does man exist? Do the world and the universe truly exist? Some thought they had the answers. The author is not so sure that there are answers.

What man can accept, what man can believe, these are issues each must decide for himself. Many atheists choose to weigh the best scientific knowledge available. They do not contend that science is an absolute truth. Science is one of several ways to view the universe. Reality, if it does exist, is an evasive commodity for which man is constantly grasping. There are times when man thinks he has found it, but soon that certainty becomes dubious.

Man now lives in a world of science. Man's achievements in technology are based upon science and its laws. For example, where would man be, and how would he live, without electricity? By these scientific laws, man regulates his life and obtains his standard of living. Only for a relatively short period in history has he lived by many of these rules. Previously, human existence was based to a great degree upon religious symbolism. Things of God were good; other things were evil. Many forms of religious symbolism have carried over into present day life, even though there have been some drastic revisions in the concepts of religion.

Before the middle of the twentieth century, some scientists believed they knew how the universe functioned. Today that certainty has become questionable. Surely one might ask if science is not just another interpretation of reality, just as unreal as man's. Is there such a thing as reality?

Many puzzles of great mystery remain in science. Some of the mysteries are concerned with the small atom and the large universe. As an example, what is light? How is it radiated from atoms? How is it transported through space? Is man living in a universe that is infinitely small and infinitely large?

At one extreme of the scientists' puzzle is the microcosm or world of the atom. This world is composed of electrons, protons, neutrons and the less common nuclear building blocks called mesons, hyperons, positrons and antiprotons. It seems only logical that if electrons have mass, they must be composed of something smaller.

The other extreme of the puzzle is the macrocosm, or universe. Our telescopes can see through the universe to galaxies of stars 500 million and more light years away. With light traveling at the speed of approximately 6 trillion miles a year, this becomes an almost inconceivable distance. Does our universe go on forever, or is this just another illusion? Einstein believed that the universe bent back upon itself like a giant bubble, having no beginning or end.

Man's ego demands answers to his questions. In past centuries, it was sufficient to answer these questions with the simple phrase, "God did it." This was the best answer in light of man's knowledge. Now man looks to science hoping to find the answers. Is not God's power gradually slipping from His hands?

The chapter entitled "Bible Contradictions" compares the Bible to present day thinking. In this chapter, only two concepts of the old religion, which were dealt severe blows by science, will be discussed. One of the Bible concepts was that the world is flat. In addition to this obvious error, the Copernican revolution theory disproved the Biblical assumption that the sun went around the Earth and could be stopped by divine intervention, as Joshua so claimed.

It was believed that the Earth was motionless and the universe revolved around it. The idea of this happening may now seem absurd to us. However, this assumption was easy to make as long as the Earth was taken as the reference point, and the fact that the other planets did not orbit the earth was ignored. When man said that something was moving, he meant the object was moving in relation to the Earth.

There is no physical experiment that can prove that the Earth is moving through space. In fact, if there were no other objects in the universe, it would be impossible to say whether the Earth was moving or standing still. Since space has no direction or boundaries, the location of an object in space cannot be defined. An object's position can only be defined in relation to other objects in space. Basically, this is Einstein's theory of relativity.

To understand relativity requires a change in one's mental picture of the universe and the world around us. When we talk in terms of the universe, it is not enough to give three dimensions for defining the location of an object in space, for in space everything is in motion. To give meaning to the location of our object, we must add a fourth dimension called time. Time is man's measure of travel in space. It establishes the location of an object at a particular moment. When the astronomer thinks of the universe, he thinks of it in terms of a space-time continuum.

There is no such thing as absolute time in the universe. Time is one of man's illusions. It does not exist without an event to mark it. Man's time is actually a measurement in space that was invented to relate events. An hour represents fifteen degrees of an arc, or one twenty-fourth of a revolution of the Earth on its axis. A year represents one orbit of the Earth about the sun. In other words, time is a measure of travel in space. Someone living on a planet that does not rotate and revolve the same as Earth would have no concept of our time.

It is believed that the laws governing one system would be the same for all systems that are moving uniformly in relation to one another. That which is experienced by one individual may be different from that experienced by another, depending upon relative positions and relative motion in the space-time continuum. The laws of various phenomena would appear almost exactly

45

the same to one individual as another. Differences in our minds and sensory organs create slight differences in observations. Since our observations are similar, we are able to establish a kind of reality in our physical world. There is always the possibility that what is apparent reality to us is not real at all. If the universe is something else from what man experiences, he has no knowledge of it. The theory of relativity is based on no other world or reality other than the one man knows. It is the physical differences in man's universe that concern relativity.

Man uses another kind of relativity in his daily thought. When he makes a statement that something is right or wrong, he is discussing it in relation to a set of personal rules that he has established in his brain. Since others will not have exactly the same rules, their statements may be different. The only world man can know is the one created for him by his senses. Everything man experiences is related to the impulses his sensory organs send to his brain. Reality to man is an interpretation of those impulses. The world as he sees it exists only in his brain.

Reality, if it does exist, is perceived by the human mind in contradictions. When man tries to comprehend reality, he is forced to look into a world that is far from that interpreted by his sensory organs. Understanding becomes difficult. Man sees colors, but actually color is an interpretation of varying wavelengths of light. Man experiences coldness and heat, which are levels of atomic activity. A cold body is one that exhibits less molecular motion than a warm one. Sound is an interpretation of the forces of moving air exerted on the eardrums. Man can see only a small portion of the spectrum of light and hear only a limited number of sound waves. His visible and audible sensory interpretation of the frequency spectrum is, indeed, limited.

Does anything in the universe exist as experienced by man outside of his mind? If there is such a thing as reality, the selectiveness of man's sensory organs may be limiting his ability to see it.

In recent years through the use of drugs, such as lysergic acid diethylamide (LSD), man has been able to cause a change in the functioning of the brain for a period of time. Under the influence of these drugs, the mind has been able to see things it had never seen before. Colors become more brilliant. The atmosphere assumes wave-like patterns. Some said that they could see the blood flowing in their veins. Nature, in general, appeared to fit into a tightly meshed pattern. Is this world of LSD closer to reality? Not likely. It is most probably just another interpretation by man of the matter around him. In some respects, these experiences may be moving man further from reality.

Does the world of man exist only in his mind, created by divine power? The nonbeliever prefers to believe that something exists, even though man's interpretation of it is distorted. He chooses to correlate the best scientific knowledge available at the time to develop his concept of the universe. When better knowledge is found, he will change his concepts.

Man's concept of God is of a non-material power, which does not occupy space, is infinite, and is present in all space. God created man, the world, and everything in it. Yet man cannot experience God. Man can only experience matter acting upon matter. For the non-believer, God becomes only an abstract word coined to designate the hidden forces of nature that lie beyond the screen of man's senses.

A basic concept of science is the hypothesis that matter can neither be created nor destroyed. All of the matter existing in the universe today has existed in one form or another for eternity. Here again, this concept could be an illusion of man's limited senses. However, until he is able to prove otherwise, it is the best concept man has. Matter exists in a number of different forms that can be transformed from one form to another. Electricity, for example, is a faster moving, less consolidated form of matter than the matter that comprises a rock.

When early man saw plants growing and objects consumed by fire, he thought of them as being created and destroyed. With his limited knowledge of nature, he did not realize that these were manifestations of matter changing form. Moreover, his life as an organized individual had a beginning and an end. It was natural for him to think of all things, including the world in which he lived, as having a beginning and an end.

To one other than a physicist, these observations may not mean much, but when they are examined by the atheist, their importance soon becomes evident. All of the forces that man experiences, such as electricity, magnetism, heat, and light, are manifestations of matter. It becomes evident that force and matter are inseparable. Force cannot exist without matter because force is a property of matter. Since matter can neither be created nor destroyed, force can neither be created nor destroyed. Since God is not composed of matter, God cannot exhibit any force and cannot have any control over the universe. This, of course, presupposes that God is not completely foreign to any concept that man may have about the functioning of the universe.

It used to be thought that electricity and magnetism were separate forces. Now, they are known to be two aspects of a single force. It appears that all forces and objects are composed of the same kinds of nuclear building blocks, and everything that man experiences is associated with these forces and objects. If this is so, then the possibility of the existence of another force, that is, the force of God, not composed of nuclear building blocks, becomes doubtful. God is then relegated only to a concept in the brain.

Let us look at some of the knowledge that science has gained about the universe before continuing with atheistic beliefs. All processes in nature show a fundamental discontinuity when they are measured with sufficient accuracy. In simpler words, no two elements in the universe are exactly alike. Scientists have found that atoms are less subject to laws than most people think. Nature's laws are more statistical averages than they are hard and fast rules.

When astronomers see an event happening on a star 100 light years away, they are actually observing what happened there 100 years ago. What is happening on that star today, they will not know until another hundred years has passed. The universe is not a fixed system. All of the solar systems are moving at different speeds and in different directions. Objects that appear to be at rest in relation to the Earth are moving in relation to the sun and other stars. Some of the galaxies in our cluster are moving toward us, while others are moving away from us at speeds up to 300 miles per second. All other clusters appear to be moving away from us. The farther away the galaxy, the faster it seems to be moving. These speeds range from 750 to 75,000 miles per second. It has been suggested that at one time, all of the matter in the universe was in one central location from whence the galaxies were born.

Astronomers, by observing the amounts of matter transformed through thermonuclear processes into radiation, have estimated the age of most of our visible solar system as being 5 to 7 billion years old.

Religions tell us that God represents the ultimate reality. If this is true, how can man know what God is or even what God is not? Does not man lack the ability to perceive reality? God is thought to be found in the act of experiencing a oneness with Him. People have said that they can feel His presence. Feelings and experiences are functions of thought. In actuality, could it not be only the concept of God with which man experiences a oneness, a concept that man gained through reading and word of mouth?

Man may be living in a world of illusions created for him by his limited perception. If man had evolved in a different manner, or had he reached a different point in his evolution, he might think differently. His observations could be different, and he might have established different natural laws to live by. The idea or concept of a God might never have entered his mind.

7. GEOLOGY AND EVOLUTION

In 1835, while visiting the Galapagos Islands, Charles Darwin observed the finches native to the islands. He noted how they had adapted in various ways to their environment. From these observations, Darwin was able to postulate his controversial Theory of Evolution. The theory stipulates, basically, that all life existing during present times has evolved from much simpler forms of life that existed in ancient times. It is easy to see how differences of opinion could develop between the evolutionists and those who believed in the teachings of the Bible. Archbishop Ussher contended that the Earth and man were created by God, 4,004 years before Christ.

Not only did Ussher stipulate what year God created the world, but he also stated the month, day and time of the day that this happened. Now that was really a trick. Also, that conflicts with the other story that it took six days for God to create everything. Then on the seventh day he rested. Now, why would a God need to rest?

Even today, with the overwhelming evidence in the theory's favor, there are people who do not believe in it. On the other hand, Darwin's Theory of Evolution is probably responsible for more people joining the ranks of the nonbeliever than any other single fact. Today we know that evolution is a fact, not a theory. You don't have to look very far today to see evolution. It is exploding all around us. Biologists can't keep up with classifying all of the new species.

Darwin noted that the finches on the islands had little competition from other types of birds. As a result, they developed into different species, adopting ways of life normal to other types of birds, had they been present. He found similar species making use of the same habitat in a different manner. He also found that the different species did not interbreed. Each species had evolved in such a way as to be best suited for its own method of obtaining food. An example is the uniquely curved beak of the species that picked worms from the bark of trees. Because of this, Darwin stipulated that species evolve by natural selection.

More specifically, the Theory of Evolution states that all plants and animals originated and descended with modification from one or a few simple ancestral forms. Species that resemble one another have probably evolved from common ancestors in relatively recent times. There often appear to be gaps in evolutionary development when one looks for fossil remnants. To assume that fossil evidence of intermediate species was not preserved is probably more accurate than to assume that later species developed suddenly from an earlier, much different ancestor. Even though there are gaps missing in the history of evolution, progression unmistakably leads from the simplest forms of life to the most complex.

Matter evolves. The matter that man calls plants and animals can be thought of as evolving in the following ways:

1. struggle for survival
2. variation of the species
3. mutation of genes
4. heredity
5. natural selection of species

One does not have to look far to see the struggle for survival among plants and animals. In recent years, it has been possible to observe how various plants and animals have adapted, in a special way, to a particular environment. It can be seen that a change in color to match the natural background has enabled one species to survive better than another. A wide range of tolerance to change in environment is an important survival factor. Plants have this greater flexibility in growth patterns. However, the behavior flexibility of animals and the ability of the animal brain to undergo modification have given the higher animals a decided advantage over plants.

To begin with, one cannot overemphasize the point that no two things in this world are exactly alike. Furthermore, variation of species is promoted by a division of a population into productively isolated groups. That is, populations living in different environments will vary in respect to the environment in which they live. There are also many variations within a group that are determined at birth. These include color of skin, eyes, and hair, bone structure and blood type.

Gene mutations can sometimes make a striking difference in an individual. Gene mutations seldom survive because the individuals must be inbred selectively to perpetuate the mutation. Inbreeding causes a decline in fertility. The mutation, itself, may cause a lowering of fitness in the individual. Animals can adjust to inbreeding, but the chance that a mutation will cause an evolutionary change is slim. On the other hand, mutations that would increase the ability of an organism to live in a new environment, or in a new way in an old environment, might foster the formation of a new species.

Through heredity, an organism receives its basic structure for life and development. By observing the development of the reproduction cells, or gametes, it is possible to see what offspring inherit from their parents. Development is also affected by the cytoplasm, or material other than the egg and sperm present in the cell. These cells are composed of chromosomes, which, in turn, contain DNA and many genes. Genes determine the heredity characteristics. There is no evidence that adaptive changes alter the genetic material of organisms. Consequently, these changes are not passed on through inheritance to the offspring. After many generations of selection, however, adaptive changes may become genetically assimilated. The human animal consists of something like 10 to 100 trillion rounded bodied cells in which the small bodies, called chromosomes, are found. There are 23 pairs of chromosomes; 23 from the male parent and 23 from the female parent. Each chromosome has many thousands of genes that determine the heredity characteristics. This mixing of chromosomes coupled with gene changes guarantees that no two people will be born the same, unless those two people are identical twins.

Natural selection is based upon the natural preference that the chemical elements have for each other and what quantity of these elements is acting upon a particular situation. Just as a farmer selects his best livestock for breeding, nature selects those best fitted to survive to sexual maturity in an environment and to reproduce. This concept was the basis of Darwin's process of natural selection. Through inheritance and natural selection, a population is constantly adjusting to its changing environment.

The time required in a laboratory to breed changes into plants and animals is short. In nature, however, the process of natural selection is much slower. Evolutionary changes are primarily due to adaptation of species to the environment and the natural selection of those species best suited for survival.

Natural selection is as effective in maintaining a population as it is in changing it, if the conditions of the population remain consistent. As this is often the case, evolutionary changes take place slowly. Usually, the average members of a population are best suited to survive within a given environment. If, however, the environment does change, the members who have characteristics better suited to the new environment will be selected to survive and reproduce. The average members of the old society will no longer be average.

There are three types of evidence that point toward evolution. The first is paleontological evidence, or that evidence gained through the study of fossils. The second is evidence found in comparative anatomy studies, comparing likenesses of structure in plants and animals. Such a study might involve men and fish. The third source of evidence is called embryology. This is the study of the development of the egg cell in animals and the seed in seed plants.

Finding fossil remains in rocks is one of the most evident records of evolution. When a seashell is found in rocks at the top of a mountain, it is quite evident that at one time in the distant past those rocks formed the bottom of a sea. It is also quite evident that it took much more than the 6,000 years since Archbishop Ussher's creation of the world for the earth's crust to be thrust upward, forming mountains.

Often, major groups of plants and animals appear suddenly in the fossil record with few or no intermediates. This could be because they originated rapidly in a limited population. However, it is more likely that, for one reason or another, intermediary fossil remains have not yet been found. The fossils of horses in the Old World show no intermediates between distinct species. This is because evolution of the horse took place in

North America. They then migrated from Alaska to Siberia during ice ages when a land bridge connected the two continents. In recent times, horses became extinct in North America. The species present today were brought over from Europe and Asia at, and after, the time of Christopher Columbus.

Also, many simple one-celled organisms that marked the beginning of life on Earth had bodies too soft to form fossils. Today, many plants and animals also fall into this category. Other early organisms may be buried too deep in the sedimentary rocks on Earth to be found. Still others that were present in sedimentary rocks may have been eroded by forces of nature. Even so, man has assembled a rather impressive set of fossils dating back over a billion years. From these fossils, the development stages of the various species can be traced. Fossils found in the deepest strata of rock were much simpler in structure than those found in recent strata. For example, the remains of man are only found in very recent strata. This shows that he has existed, as we know him, for probably not more than 5 million years. It should be noted that during the lower Cambrian Period, there were about 500 species of living organisms in existence. Today, there are millions and millions of species. This increase from the one-celled critters such as blue-green algae occurred over a period of approximately 600 million years.

One might ask how geologists know how old rocks are. There are a number of ways in which rocks can be dated. One widely used method is the measurement of the disintegration of radioactive isotopes found in rocks. In 1896, a physicist by the name of Antoine Henri Becquerel discovered that certain elements having unstable atomic nuclei, like uranium, spontaneously disintegrate at a constant rate to form more stable products, such as lead. This rate of disintegration can be very slow, and it can be measured with a high degree of accuracy. Changes in physical and chemical environments have no apparent effects upon the disintegration rate. Small quantities of unstable elements are found in most igneous rocks. Igneous rocks are those formed from molten materials. By measuring the proportion of lead to uranium, the age of a rock can be calculated.

By correlating studies of the various radiogenic isotopes found in rocks, a fairly accurate estimate of age can be determined. Often, molten or igneous material has been intruded into sedimentary beds containing fossils. Geologists can determine the age of the igneous intrusion. They also know that the fossils in the sedimentary beds are older than the intrusion. The oldest rocks found have been estimated to be 4.6 billion years old.

Now let's talk about radioactive isotopes. There is quite a large number of them. Some have a half-life of only seconds. For the geologist they are useless. Others have half-lives of over 100 billion years. Herein I have included a list of some of the more common ones used by geologists.

Radioactive Isotope	Decays to	Half-life ($t_{1/2}$)
Carbon-14 ^{14}C	Nitrogen-14 ^{14}N	5730 years
Beryllium-10 ^{10}Be	Boron-10 ^{10}B	1.5 million years
Uranium-235 ^{235}U	Lead-207 ^{207}Pb	704 million years
Uranium-238 ^{238}U	Lead-206 ^{206}Pb	4.4 billion years
Potassium-40 ^{40}K	Argon-40 ^{40}Ar	11.93 billion years
Thorium-232 ^{232}Th	Lead-208 ^{208}Pb	14.01 billion years
Lutecium-176 ^{176}Lu	Hafnium-176 ^{176}Hf	35.7 billion years
Rhenium-187 ^{187}Re	Osmium-187 ^{187}Os	42.3 billion years
Rubidium-87 ^{87}Rb	Strontium-87 ^{87}Sr	48.8 billion years
Samarium-147 ^{147}Sm	Neodymium-143 ^{143}Nd	106 billion years

Now what is a half-life? It is the time when half of the radioactive isotope has decayed. The next half-life is when half of the remaining isotope has decayed. The rate of decay is called exponential. In other words, you have to be a mathematician to calculate the age of a rock. But when does this rock begin to decay? Year "0" is when it has solidified from molten magma. At that point there is only the radioactive isotope present. There is no decayed material. Now you will note that I have listed carbon-14. Carbon-14 does not usually deal with rocks; it deals with plants and animals. Year "0" happens when the plants and animals die. At that point, they are removed from the food chain. The amount of carbon-12 and carbon-14 does not change after death. There is far less carbon-14. Carbon-12 is not a radioactive isotope, so its quantity is fixed. As carbon-14 decays, the remaining carbon-14 is compared to the carbon-12, whereby the age can be determined. In most cases carbon-14 dating is useless for geologists because of its short half-life.

It has been only since the mid-1900s that scientists have developed instruments capable of analyzing very small quantities of radioactive rocks and other materials. One may remember the story about the Shroud of Turin. It was thought that it might have covered Jesus after he was crucified. Recently, the Vatican, which is its present custodian, allowed tiny pieces of the shroud to be removed and analyzed by three prominent laboratories. Their results ranged from about 1200 to 1300 AD, so the shroud still remains a mystery.

There are many theories on how the earth was formed. Some believe that it consolidated from a molten form. It has been estimated that this occurred about 5 billion years ago. Others believe that it was formed from interstellar debris. Whatever the origin may be, the Earth's crust seems to have become stable enough to support life 3.5 billion or more years ago.

How did life begin? No one can say for certain. But we do know, if the dating of rocks is accurate, that there has been life on this planet for a long time. Blue-green algae have been found in the Fig Tree Formation of South Africa, which is dated at 3 to 3.5 billion years ago. Calcareous algal reefs, dated about 3 billion years ago, have been found in the Bulawayo strata of Rhodesia.

It is thought that bacteria and viruses existed before the single cells, such as blue-green algae.

Some have postulated that at first the Earth most likely was very much hotter than it is today. The atmosphere was mainly composed of hydrogen, helium and lesser amounts of hydrogen sulfide, methane, ammonia and water vapor. Many of these early hot gasses escaped into space. As the earth began cooling, water vapor, nitrogen and carbon dioxide were released from the crust into the atmosphere. The water vapor condensed to form great seas.

To find how life might have started under these early conditions, several scientists have conducted experiments. Stanley Miller, in 1953, showed that when methane, ammonia, hydrogen and water were subjected to an electrical charge, numerous complex organic molecules were produced. These elements were probably present in large quantities in the Earth's early atmosphere. The molecules produced included protein and amino acids necessary for the formation of life. Also, Sidney Fox in the 1960s produced protocells that had a membrane which is essential before life can be said to exist. But these cells did not live for long, and their function was very limited.

Salts and other compounds were washed by the rains from the land into the sea. Cosmic rays and ultraviolet rays bombarded the atmosphere, altering elements and forming organic compounds. The sea became an organic sea, where colloidal clusters separated by a thin film of water began to form. The colloidal clusters were in a continual state of being built or broken down. Those that could attract more molecules than they lost, grew. Some of them absorbed molecules which acted as catalysts in improving growth. The catalysts later developed into the present day enzymes that trigger cell activity. If a cell grew too large, it split in half, increasing its absorptive surface.

Even in the early stages, one can see the effects of natural selection. It was the survival of the fittest. Those cells best suited to grow divided and multiplied. Others starved and died.

After some time, organic material in the seas became depleted, and a struggle for existence among the colloidal clusters ensued. Some colloidal clusters with a favorable chemical composition were able to convert simpler organic compounds into the more complex ones that they needed for food.

As food became scarce, some cells developed the ability to make organic food from inorganic compounds by photosynthesis. Here, again, is natural selection and evolution at work. Through photosynthesis, oxygen was released to the atmosphere. Some cells were able to use the oxygen in burning food. Simple cells banded together forming complex organisms. Each cell had a specific function in enabling the organism to succeed over other organisms.

When oxygen was released to the atmosphere, the ultraviolet rays from the sun converted the Earth's upper atmosphere into ozone. This ozone layer, in turn, filtered out many of the life-killing ultraviolet rays, thereby making it possible for life to move from the sea to the land.

All living things are made up of cells. The simplest forms of life consist of a single cell. The larger plants and animals consist of many cells, which are combined into different groups, each doing a special job to keep the body functioning. Evolutionary changes have fostered the development of more and more complex organisms in terms of different types of parts comprising the whole. All of today's complex plants and animals are descendants from the simple one-celled organisms that lived in the seas of the world over 3 billion years ago.

The best evidence in support of evolution lies in the marked similarities that plants and animals exhibit from one family to the next. It is not hard to believe that they descended from common ancestors. Striking similarities in bone structure exist among many animals. Bird's wings and man's arms have a single bone in the upper arm and two parallel bones in the forearm.

[SEE EVOLUTION OF LIFE ILLUSTRATION]

Dates of geologic periods and evolutionary emergence are approximate.

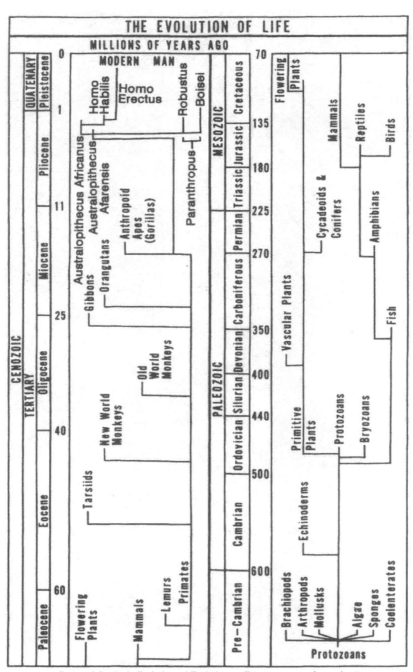

The Evolution of Life

Dates of Geologic Periods and Evolutionary Emergence are Approximate

60

In addition to what is shown in the Evolution of Life chart I want to add some geologic times of other critters that are not shown.

Life Forms	Millions of Years BC
Dinosaurs	65 - 228
Trilobites	225 - 570
Jellyfish	- 1050
Red alga	950 - 1260
Protozoans	- 1750
Eukaryotes	- 2600
Prokaryotes	- 3500
Stromatolites (blue-green algae: no nucleus)	- 3500

Another assumption, of course, is that God made plants and animals in a similar way. The nonbelievers find this assumption hard to accept. Also, plants and animals go back many millions of years, while gods were invented by man less than 100,000 years ago.

Other evolutionary evidence can be found in the study of the embryo or egg. It is virtually impossible to distinguish between an early embryonic sheep and an early embryonic man. Another striking example is found in the fact that both birds and mammals possess gill pouches in the embryonic form. These develop into parts of the ear, throat, tonsils and various glands before birth. In the case of fish, from which birds and mammals are believed to have descended, the pouches still develop into gill slits. Such embryonic forms may persist primarily because they play an important part in a long chain of reactions that eventually lead to final development. In other words, the embryo grows in relation to the evolution of the species.

Because of the many similarities that exist between plants and animals, it is possible to classify them into families. By doing so, one is also reflecting evolutionary relationships. By using this type of classification, it is easy to see how one species evolved from another or how several species evolved from a common ancestor.

In 1828, urea, an organic compound, was synthesized in the laboratory from inorganic compounds. Urea, an end product of protein decomposition, is used in the synthesis of resins and plastics and in fertilizers. Since that time, many other organic compounds have been made in the laboratory. They include alcohol, ether, grape sugar, racemic acid, oxalic acid, formic acid, butyric acid, acetic acid, lactic acid, fat, amyloids and alkaloids. Dr. Sidney Fox has been able to produce simple proteins held together by double-walled spherical cell-like envelopes. These proteins (proteinoids) multiply by division, as do cells. Man today can make amino acids and from them, synthesize DNA. If man can do it, why couldn't nature develop living cells billions of years ago, given the right conditions? By manufacturing life in the laboratory, man disproves the contention that only God can create life.

The gap between living cells and inorganic matter, such as rocks, is filled by inorganic crystalline cells called crystalloids. These cells exhibit many of the same properties of protoplasm. Also, there is no fine line between plants and animals. Some forms of life are almost halfway between the two.

Many plants and a few animals reproduce by subdivision. At first, this process seems superior to sexual reproduction, since having two sexes greatly decreases the potential rate of population increase. But asexual reproduction lacks evolutionary plasticity and in the long run may lead to population extinction. The sexual process, on the other hand, increases the range of potential variation by bringing together genetic changes in different lines of descent. Sexual reproduction may have evolved from one single-celled organism swallowing another. Part of the chromosomal material of the swallowed cell was not digested and persisted, becoming part of the cannibalizing cell.

Normally, members of unrelated species will not mate. If they do, the first generation may die or exhibit genetic weaknesses. Consequently, they contribute little to future generations.

After the egg has been fertilized, it divides into two parts, then into four parts, and finally into many millions of different cells, each performing a specific function within the organism. The initial cells appear to have the same genetic construction. As development progresses, the cells exhibit different specializations, which are determined by their position in the embryo and by the exerting influence of the neighboring cells.

One species that has existed for 300 million years is the lungfish. They adapted to a stagnant-water environment. Two existing species have survived by burying themselves in mud when their pools of water completely dried up. As is common with other major groups, the lungfish evolved rapidly, soon after their origin. Over the long period that followed, they changed very little. The osteolepis fish evolved in such a way that when one body of water dried up, they could travel over land to another. It is from this group of fish that the amphibians, and later, land vertebrates evolved. Contrary to popular belief, evidence shows that man has not descended directly from monkeys and apes. They have all descended from a common ancestor.

One of the earliest known creatures resembling a man is Australopithecus Afarensis, who lived in Africa about 4 million years ago. Its teeth and skulls resembled man's, and the construction of its pelvis and thighbones enabled it to walk erectly. The earliest tools were made about 2 million years ago by Paranthropus Robustus. But it was Australopithecus Africanus who survived to give rise to Homo habilis and eventually to Homo erectus. Homo erectus began spreading to other parts of the world about 1.8 million years ago. Some believe that modern man (Homo sapiens) developed in Africa and then spread to other parts of the world beginning about 100,000 years ago. Eventually, Homo erectus died off around the world and Homo sapiens replaced him.

Modern man has many rudimentary organs, such as the appendix, the tonsils, the ear muscles and the semi-lunar of the eye. He no longer has a need for these organs that were left over from a past period in his evolution.

The composition of the human body is constantly changing. Even more solid parts of the body, like bones, do not contain the same molecules at age 50 that they did at age 30. The molecules remain in the cells for only a short time. If the composition of the cells is changing, the cells themselves are changing, even though they maintain, to a degree, their initial patterns. Through this process of change, man experiences aging.

Modern man has an advantage over other forms of life. He does not have to wait for genetic assimilation of new adaptive advances. Such advances are passed on by cultural, rather than genetic, means. Man can change his environment to suit himself, rather than develop a new genetic type to fit a new environment. As an example, man has the ability to live in space. The last several hundred years have witnessed a great acceleration in the rate of man's technical progress. This is because the first steps are the more difficult, not because man's intelligence has greatly increased. Early man was too busy searching out a living and fighting his enemies to do more than copy patterns handed down to him by his ancestors.

Almost all species existing in ancient times have eventually become extinct. In recent years, many species have become extinct due to man's activities. Man has changed the environment of the world to some degree. Consequently, plants and animals have had to become better adapted to this new environment. A common occurrence is to find that the coloring of a species has changed to blend with the environment that man has created.

Needless to say, in the mind of the evolutionist, death and disease have existed since the beginning of life. Death and disease were not created at the time of Adam and Eve as punishment for sinning. To the nonbeliever, love, reason and justice are not God given. They have evolved through the ages because they have been advantageous to the development of man in his environment. The only meaning in life is that which man, himself, gives to it. There is no supernatural power to give him guidance.

Now let's get back to geology. It was at the beginning of the 20th century that a German by the name of Alfred Wegener came up with the theory of continental drift. As an example, before that time it was thought that Antarctica had been more tropical because explorers had found coal there. Wegener noted that the rock formations on the west coast of Africa matched perfectly with the rock formations on the east coast of South America. By the way, we no longer call it continental drift. We now refer to these moving land masses as tectonic plates. The whole surface of the Earth is covered with these tectonic plates. Some are very large, such as the Pacific plate, and others are small, like the Juan de Fuca plate. The crust of the world is about ten miles thick. Of course, it is thicker under mountains and may be only two miles thick in some other areas, such as where volcanos are present.

At one time in the distant geological past the world had two super continents. One was called Gondwanaland. Keep in mind that the molten magma in the center of the Earth flows convectionally, and it exerts a force on the tectonic plates that ride upon it.

Now about 165 million years ago Gondwanaland began to break up. South America began slowly moving to the west. Antarctica, India, Australia and Madagascar together started moving to the east. Let me reiterate that these movements took millions of years. Eventually Antarctica split off and headed towards the South Pole. The Indian plate moved over and butted up against the Eurasian plate, causing the Himalayan mountains to raise. Australia went where it is now, and Madagascar stayed close to Africa. Also, you must keep in mind that the plates are still moving.

Richard Dawkins believes that some day (maybe many millions of years from now) the part of California west of the San Andreas fault plus Baja, California, will become an island. What is happening is that the Pacific plate is rubbing up against the North American plate, rotating northward at about seven inches a year. However, I am not so sure that western California will ever become an island. It is easy to see what happened in the past and what is happening at present, but considering the multitude of factors that influence planet Earth, I think that it is impossible to predict what will happen millions of years form now. It is too bad, but neither of us will be around to see who was right.

8. GENETICS, DNA, RNA, ETC.

All living things, are made up of cells. The first single-celled creatures that existed 3.5 billion and more years ago had no nucleus. Life moved very slowly in those days. It took millions and millions of years for cells with a nucleus to evolve. Then, finally, maybe about 700 million years ago, cells began to combine to make more complex organisms. By the way, some believe that viruses and bacteria existed before other cells because they are simpler forms of life.

Now you know that it took a long, long time for life to get as complex as it is today. In fact, it is so complex today, that we are only beginning to understand it. Before we get into the structure of cells, their DNA, genes, etc., let's review some of the things that have already been said.

Without a brain, thought and memory become impossible. For example, rocks have no knowledge of their existence. Yet, they are made of the same nuclear building blocks, that is, electrons, atoms, etc., as life. Remember, atoms, molecules, cells, etc. have no knowledge of their existence because they do not have brains. They act automatically the same as they have for millions and billions of years.

It is a scientific fact that the human brain can only receive material inputs, such as photons of light hitting the million or so light sensitive receptors in the retinas of our eyes, thereby generating electrical impulses to our brains. Even if spirits had a sub-molecular structure that man's instruments could not detect, spirits would still have no knowledge of their existence, since matter, other than that of brain cells acting together, has no knowledge of its existence. A rock has no knowledge of its existence. An atom has no knowledge of its existence. A spirit, if there were such a thing, would also have no knowledge of its existence. Again, spirits cannot think because they have no brain cells to form a memory and develop thoughts. When the brain dies, the lights go out, and thought becomes impossible. There is nothing left to produce thought. A nonmaterial spirit cannot have the capability of thought. If only matter can act upon matter, a nonmaterial god cannot build a mountain. Only the tectonic forces under the crust of the earth can do that. An example would be the Indian plate pushing up against the Eurasian plate, thereby causing the Himalayas to rise.

Please note that the brain cells of the higher animals are far different from the brain cells, or their equivalent, of the lower animals, such as sea snails. Plants also have a cellular structure, though different from animals, and their growth, production of seeds, and death are determined by their genetics.

One of the objections that people have about science is that it is always changing. They like religion because it is simple and appears not to change. However, the reason why science is changing is because everyday we are learning more about it. In fact, most of our knowledge has been achieved in the past 50 years or so. For example, we used to believe that Antarctica was tropical in past years. Now we know that continents drift, and about 40 million years ago, Antarctica arrived from a more tropical location to about where it is located now. At one time, Antarctica was joined with Australia to South America, Africa and India, forming the super continent of Gondwana.

Please note that nowadays scientists call nuclear building blocks by different names than I have used in this book. However, no matter what their names are, they are still the same.

To understand how genetics and DNA evolved, it is necessary to understand how the Earth evolved and how scientists have come about to understand its evolution. Let's begin with the origin of the Earth.

Geophysicists have estimated that the Earth formed about 4.6 billion years ago by the consolidation of space debris, that is, by the coming together of cosmic dust, gravel, balls, etc. No life was possible on Earth at that time because of constant volcanic eruptions and bombardment from space by planetesimals. Geophysicists determined the age of the world by examining the oldest rocks that they could find. Determining the age of rocks is done by measuring the rate of decay of one radioactive isotope into another material. That is, uranium 238 into lead 206, or uranium 235 into lead 207. There are many radioactive isotopes.

See the chapter on geology for a list of some of the most common ones used by geologists.

When one thinks about it, if there wasn't evolution, it would be next to impossible for universal features, like eyes, ears, noses, toes, etc. to have evolved separately. They had to be inherited species to species from some early ancestor. Also, the genetics of species are almost identical. The proteins in modern organisms are all fashioned from one set of standard amino acids.

By studying genetic data, it is now believed that modern man has evolved only very recently. Perhaps 100,000 or less years ago, he evolved in Africa and then spread gradually to the rest of the world. At this point, I want to reiterate that exact dates are not important to the atheist. What is important is that it didn't happen a few thousand years ago.

Going back 7,000 generations, the more than 7 billion people in the world today evolved from 60,000 people. All people are genetically 99.8% identical. Yet, we kill each other for being different, and that difference is often religion. As you will see, it only takes that 0.2% to make us all different, that is except for identical twins.

Our closest relatives in the animal kingdom are chimpanzees. Their genetic makeup is 98.5% identical to that of humans. However, their brains are considerably smaller. It is also interesting to note that some worms have genetics that are 75% the same as humans. We are also close to mice (2.3 billion letters), birds and reptiles. Our common ancestor (maybe a lemur) lived 8 million years ago. 7 million years ago, the gorilla line split off, and 5 million years ago, humans and chimps split off into distinct species.

It is interesting to note that Charles Darwin was subjected to a great deal of criticism during his lifetime. His theory of evolution was rejected by almost everyone. The idea of "natural selection" occurring due to changes in the environment was more than people could accept. In the end, he appeased them by saying that the "Creator" originally breathed life into one or a few life forms.

Is life something special, or is it just another form of matter? There are many forms of matter. Under the right conditions, light will be produced, and under the right conditions, life will be produced. It is quite probable that there are forms of matter that do not exist on Earth but do exist elsewhere in the universe. Life is just another form of matter. There is no special power that causes it to exist.

[SEE CHROMOSOMES, DNA, ETC. ILLUSTRATIONS]

[SEE DNA AND RNA STRANDS ILLUSTRATION]

Now let us look at what we know about genetics and the human genome. First I want to point out that chromosomes are more X shaped and not cylindrical, as pictured in "Cells, Chromosomes, DNA, Etc."

Let's start out with the human body. It is made up roughly of 50 to 100 trillion cells. Of course, small people have less, and obese people have more. Each cell has a nucleus somewhere near its center that is surrounded by the cytoplasm, which is enclosed by a flexible cell membrane. It is inside the nucleus were the chromosomes and DNA (deoxyribonucleic acid) are found. There are 46 chromosomes with 23 coming from the father and 23 coming from the mother. Each chromosome is composed of some protein with the DNA twisted into packets around it.

Now the total DNA is composed of about 3 billion nucleotides in each cell. For simplicity let's call them bases and base pairs. They are "A" (adenine), "T" (thymine), "C" (cytosine), and "G" (guanine). Also for simplicity let's only use the letters "A", "T", "G" and "C". "A" is always paired with "T", and "G" is always paired with "C".

Now let's pause a minute to talk about genes. It has been estimated that there is between 20,000 and 25,000 genes in each cell. Actually, genes are nonmaterial. They are only names given to lengths of DNA that code for a particular protein. On an average, there are about 50,000 base pairs for each gene, but in actuality nothing is average. The number of base pairs per gene can vary greatly.

Of the 3 billion bases only about 3% code for proteins. What happened to the other 97%? I think that it is a good possibility

CELLS, CHROMOSOMES, DNA, ETC.

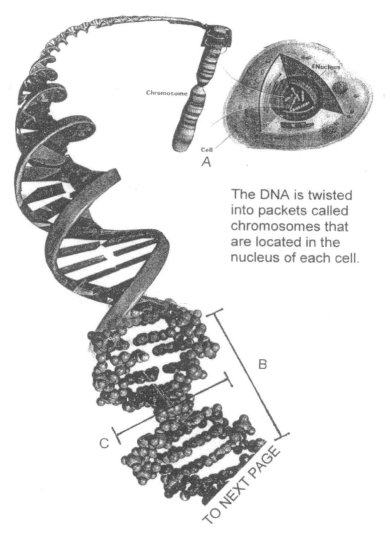

Chromosome

Nucleus

Cell

A

The DNA is twisted into packets called chromosomes that are located in the nucleus of each cell.

B

C

TO NEXT PAGE

B = length of one turn of DNA is 34 angstroms
C = diameter of DNA is 20 angstroms

1 micrometer = 0.00003937 of an inch
1 angstrom = 0.000000003937 of an inch

CELLS, CHROMOSOMES, DNA, ETC.
(continued)

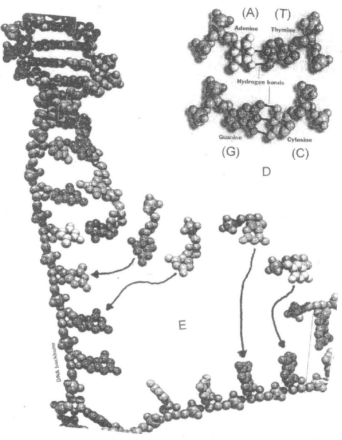

D = The blueprints for making proteins and
building new cells are stretches of DNA called
genes. They have been given the letters (A),
(T), (G) and (C)—(A) bonding to (T) and (G)
bonding to (C) with hydrogen bonds.

E = To copy itself, the DNA unzips along its length
forming two reverse image half ladders, then
each half rebuilds itself with components from
the cell forming two identical DNAs with (A)
bonding to (T) and (G) bonding to (C).

DNA AND RNA STRANDS

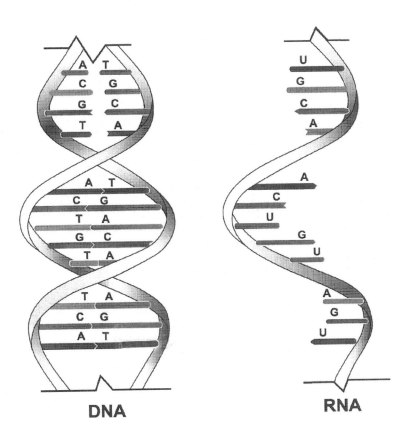

DNA

RNA

These are an artist's concept of DNA and RNA. They may be far from real, but they do show how the DNA unzips down the middle and produces a mirror image of itself. Then messenger RNA carries the copies out of the nucleus to ribosomes in the cytoplasm where transfer RNA gathers up the amino acids necessary to make a protein. This process could involve just making several genes, or it could involve making a whole new cell.

that over the past millions and billions of years of evolution they did code for proteins, but as the structures of bodies changed they were no longer needed and they became dormant.

Now I have to ask myself why go on about how cells reproduce themselves. I have produced overwhelming evidence that the process is extremely complicated, and no man-made god could have anything to do with it.

But I will say in a very simplified way that at first the DNA has to unzip itself down the middle and make a copy of itself. Then messenger RNA brings that code out of the nucleus to the ribosomes in the cytoplasm where transfer RNAs assemble the amino acids required to build a new cell. The ribosomes are the cells' workbenches. If only one or more proteins are needed, only that portion of the DNA representing those proteins will unzip and make a copy. If you have ever seen an electron microscope's view of the cytoplasm you will see that it is packed with many other parts that get into the act.

By the way, there are only 20 amino acids that produce the thousands of proteins that are in all of the life forms without any aid from any gods. Also, both DNA and RNA are made up of one kind of molecule called nucleic acid, but their end products are composed of a very different molecule, which is a protein. Also note that the "T" in DNA becomes "U" (uracil) in RNA.

Also, organisms as different as humans and bacteria use the exact same coding rules. Recently, scientists have found at least sixteen variants to the code, but, in general, the code has existed in all life forms for billions of years. From time to time, other codes existed, but in the end, the standard genetic code won in the struggle for existence.

9. PLANT GENETICS

As previously stated, all living things are made up of cells. Some are nearly round; some are disc-shaped; some are long and skinny; some are rectangular; etc., etc. We think mostly about animals and plants, but there are also three other groups of living things. They are protists (algae), Monera and fungi. I might mention that the fungi group includes yeasts, molds and mushrooms.

Most protists are one-celled critters, like diatoms. However, the algae we think of the most is the giant kelp waving in the ocean and growing to be well over 100 feet long.

Back to plants. They can be made up of many more trillions of cells than man. The giant redwoods would be a good example.

Like animal cells, plants have a nucleus and a cytoplasm. The cytoplasm is the area between the nucleus and the cell membrane. The cytoplasm has many of the same working parts as the animal cytoplasm. There are some differences. Plants make their fuel form water, carbon dioxide gas and sunlight in their chloroplasts. Their fuel is a sugar. Oxygen is their waste product. Animals get their fuel from eating plants or other animals.

There is usually only a single vacuole (water storage area) in plant cells, while there are usually several in animal cells. When you see a plant's leaves drooping it is because its vacuole is running out of water.

Another big difference is that animals have only one flexible cell wall. Plants have a flexible inner cell wall and a fibrous outer call wall. The outer cell wall is composed of cellulose, which they make form sugar. The cellulose is what gives wood, cotton, etc. its strength. Plants also make proteins which animals eat.

Most plants are flowering plants. The flower has both male (stamen) and female (pistil) parts. The plants depend on bees, hummingbirds and other critters to pollenate their flowers. In exchange the critters get nectar to eat. Wind also can pollenate flowers.

Before going on to genetically altering plants, I want to regress to animals again. It was originally thought that some dogs were bred from jackals, but since the invention of the electron microscope in the middle of the last century and the discovery of DNA, it has definitely been proven that all dogs have been bred from wolves. By the way, the electron microscope can magnify up to 200,000 times. The best of ordinary microscopes can only magnify 2,000 times.

Gregor Mendel (1822 - 1884), an Austrian monk, is accredited as being the first to crossbreed plants. He bred and crossbred pea plants and observed the differences that his crossbreeding produced. He published his findings in 1866, but at that time no one was interested in his findings.

Since the beginning of the 20th century there have been many kinds of crossbreeding done. One example would be taking a plant that is resistant to disease but tastes bad and crossbreeding it with a plant that tastes good but is not resistant to disease in order to produce a plant that is resistant to disease and tastes good. Another example would be grafting parts of one plant onto another plant, thereby producing a more desirable plant.

Nowadays botanists use a new technique called genetic engineering. This technique involves inserting genes into the sex cells of plants. Many people oppose this technique because it is unnatural and could produce unhealthy foods.

If you have ever seen wild cabbage you would probably call it a weed. But when that weed was genetically engineered they came up with all kinds of plants. The list includes white cabbage, red cabbage and all the other cabbages, broccoli, cauliflower, broccoflower, kale, kohlrabi, Brussels sprouts, etc.

Another type of genetic engineering is what they did to the tomato. They thickened the skin and lowered the acid content so that it would travel better. Of course, that destroyed the original taste.

Man did not only do all of these alterations, but he also invented God to explain the unexplainable. The only problem is that God only exists in the mind of man.

10. SCIENCE, THE BRAIN AND HOW IT WORKS

The last 10 years of the 20th century have been called "the decade of the brain." More has been learned about the brain in that time than in previous times. Science, in general, has made great advances in the past 50 to 100 years. This chapter will also present the theories that have evolved in the last few years.

First we will talk about the general aspects of the brain. Then we will talk about how ideas concerning the brain have changed.

It can be said that "reality is in the eye of the observer," or more accurately, "reality is in the brain of the observer." We don't see at the eyes, smell at the nose, hear at the ears or feel at the fingertips. All of these functions are accomplished in the brain.

Because people are constructed basically the same, they perceive the matter that composes the world basically the same. However, there are differences that exist from person to person, and consequently, everyone perceives matter a little differently.

Now, when I talk about matter, I am talking about everything in the universe from the least compact substances, such as photons of light, electrical charges, and gases, to the more solid substances, such as granitic rocks and diamonds.

By the way, most people do not realize that a photon of light has the equivalent mass of 1,047 electrons. Light is composed of matter.

We are not aware of all the forms of matter. Without instruments to help us, our knowledge would be very limited. For example, without a radio, we would not be able to receive radio waves. It is also possible that there are forms of matter yet to be discovered.

It has taken hundreds of millions of years of evolution for the human brain to develop to the point where it can grasp whole scientific concepts. At the same time, we must remember that it has only been the last few thousand years where the most progress has been made. Before present times, man lived in accordance with superstitions. As a consequence, even today the human brain is genetically structured to accept these superstitions. Also, remember that it was only recently that the Bible was written. At that time, man knew little about his surroundings. He thought that the world was flat and that God created the world in the year 4004 B.C. or, at least, that is what Archbishop Ussher postulated. Even today, many people find science too overwhelming to understand. When I talk about science many of my friends say, "Oh, I am not interested in that."

As science has moved forward, the gap has widened between what we are capable of observing and what scientific instruments are capable of observing. These instruments perceive many more aspects of the universe than can the human senses. Scientific findings can also be measured with far greater accuracy than is humanly possible. One of the more recent developments of science is the basic concept of how the human brain functions. Of course, the specifics need much more work. However, even the most basic brain functions tend to be ignored by most people. Highly religious people refuse to believe them at all. Some even believe that feelings come from the heart.

The question arises, does anything in the world exist as man perceives it? Atheists tend to weigh the best scientific knowledge available to establish a reality. They do not contend that science will lead us to an absolute truth. Science may be just another way to view the universe. Remember, as we will learn later, the brain can only function in relation to the inputs that it receives. It cannot store information for which it has not received an input from the sensory organs or the genes that have constructed it; but, on the other hand, it can generate ideas based upon the inputs it receives. One example of this would be the generation of the idea of God to explain the unexplainable.

The atheist believes that man's knowledge of God has been developed as an idea in his brain and taught from generation to generation. Man cannot otherwise perceive God because God is nonmaterial and cannot act upon any of the body's sensory organs. That is, man cannot see God or hear God or feel God. Only the concept of God has been stored in and is recalled from his brain.

Again, if a material body driven by a material brain can accomplish all of the functions that it does, why should it need a nonmaterial spirit to reside in it? It doesn't. Only matter can act upon matter, and we can only perceive matter, not nothingness. The only functions of the idea of spirits are to create a father image to guide us, to explain the unexplainable and to make us not feel so bad about dying. We are told that our spirit will go on to a better world someplace in the sky. The only problem is that this someplace has no specific location, and nobody knows much about it. To the author, that sounds like a fiction of the imagination of a man's material brain. Also, if spirits were material, that is, composed of material building blocks, the instruments that man has developed to detect those building blocks would be able to sense spirits as well. It should be possible to magnify and show them on a television screen. Unfortunately, this is not the case. Again, we come back to the conclusion that the spirits are only ideas stored by material brain cells in the mind of man.

Man's ego demands answers to his questions. In past centuries, it was sufficient to answer these questions with the simple phrase, "God did it." This was the best answer in light of man's knowledge. Now we look to science hoping to find the answers. Today, it is hard to believe the biblical assumption that the sun went around the Earth, and it could be stopped by a divine intervention, as Joshua so claimed. Also, many years ago, the Bible said that the world is flat. That was one of the many writings that were stricken from the Bible as the years passed.

It is only natural that early man should fabricate gods to explain the unexplainable. His points of reference were far different than today. His knowledge of the universe was shaded by his limited ability to observe it.

Over the years, as the number of scientific discoveries increased, it became evident to the nonbeliever that the world was not regulated by a divine power. All matter present in the world exhibits power. What happens in all the universe is due to the action and interaction of the material building blocks comprising the universe. If this is true, then the idea that God controls the universe is not true. God is then relegated to being only a concept stored in the brain of man. God cannot be experienced otherwise because God is not matter and cannot act upon any of man's senses. For the nonbeliever, God becomes only an abstract word coined to designate the hidden forces of nature that lie beyond the screen of man's senses.

In one sense, all matter is living. The nuclear building blocks that compose a rock, when observed under an electron microscope, are seen to be actively moving. It might be true that organic compounds, such as plants and animals, exhibit more molecular motion than inorganic ones. Nevertheless, all matter is molecularly in motion and probably has been molecularly in motion forever. The mass or weight of a system varies with its motion. It increases with velocity relative to an observer. Motion is a form of energy. Increased mass of a moving body comes from its increased energy. Energy has mass. In a sense, solid matter is concentrated energy. All of the forces that man experiences, such as electricity, magnetism, heat, and light, are manifestations of matter. It becomes evident that force and matter are inseparable. Force cannot exist without matter because force is a property of matter. Since matter can neither be created nor destroyed, force can neither be created nor destroyed. Since God is not composed of matter, God cannot exhibit any force and cannot have any control over the universe.

Again, it appears that all forces and objects are composed of the same kinds of nuclear building blocks, and everything that man experiences is associated with these forces and objects. If this is so, then the possibility of the existence of another force, that is, the force of God not composed of nuclear building blocks, becomes doubtful. God is then relegated to be an idea, stored by the brain cells in man's brain.

If we had evolved in a different way, our brains may have been differently constructed. Our observations would be different, and we might have established different laws to live by. The idea of a god regulating the universe might never have entered our brains. In fact, man is probably the only animal that has a belief in gods.

We experience coldness and heat, which are levels of atomic activity. A cold body is one that exhibits less molecular motion than a warm one. We can see only a small portion of the spectrum of light and hear only a limited range of sound waves.

Do things exist in the universe as experienced in our brains? If there is such a thing as reality, the selectiveness of our sensory organs and construction of our brains may be limiting our ability to perceive that reality.

Consciousness or awareness is a function of the brain that is hard to explain. It used to be thought that consciousness was a nonmaterial spirit called the mind, which resided in the brain. But now when we look at the brain, we see that everything that it is composed of and functions by has a material composition. It does not need a spirit for it to function. The exact process of awareness is unknown, but it has been suggested that the firing of some of the brain cells symbolize awareness. A model representing space outside the brain is created inside the brain.

Because the information is continuously coming into the brain at a rapid pace, an illusion is created that the seeing is out beyond the eyes, the hearing is out beyond the ears, and the feeling is at the fingertips instead of the brain.

Again, we live in a material world. Everything we sense is composed of material building blocks. Light, heat, gases, liquids and solids are all composed of the same building blocks. That is, electrons, protons, etc. form atoms, which form molecules, which form elements and the material that we can sense. Since we cannot sense nonmaterial things, such as gods, spirits and the hereafter, these ideas become relegated to material cell structure within the brain.

Let us remember that in the name of the spirits, whom atheists do not believe exist, kings and religious leaders have subjected their parishioners to a form of slavery to escape eternal Hell. In reality, Hell most likely does not exist. It is the goal of the atheist to free the religiously oppressed from the slavery of religion. We have no desire to convert those that need religion to live by and to die by. It is also the desire of many atheists to preserve the separation of church and state so that everyone can believe in what they want within the boundaries of a country's laws.

Next, let us compare the brains of man (Homo sapiens) and other animals. Sea snails have only about 20,000 brain cells (neurons). Their ability to function is very limited. They can advance to food or retreat from danger. I feel certain that there is no place in their brains for storing ideas of gods, spirits or a hereafter.

Again, it is dubious whether any other animal besides man has any idea of gods, spirits or a hereafter. Man's unique brain structure goes far beyond that of almost all animals with its ability to form words and create ideas within the brain.

When one thinks of the functioning of the brain in terms of the transmission of electrical impulses coming to it, it seems evident that we do not see at the eye or hear at the ear. Rather, the realization of sight and sound are accomplished in the brain. We do not feel pain at the surface of the skin. The pain is actually felt in our brains. It is only an illusion that those parts affected are actually feeling pain. Sever the sensory nerves to them, and there will be no feeling of pain.

Thinking in terms of sensory receptors sending signals or electrical impulses to the brain, it seems evident that there is no such thing as a thought whose subject was not received by the sensory organs at one time or another.

However, thoughts can be processed by the brain in such a way as to give the impression that the resultant thought was generated by itself.

We are conscious of our will but ignorant of the causes that will induce it. As already implied, we have no real free will.

Our actions are governed by the interaction of sensory inputs with our established memory processes. These memory processes are continually undergoing change as a result of the influence of new sensory inputs. Reactions are not caused by one input. Often hundreds of thousands of inputs are involved in a reaction. The numbers and types greatly vary. Consequently, one will not get exactly the same reaction from what appears to be the same input. Due to this complexity of action, man receives the illusion that he has free will. In addition, it soon becomes evident that brain and mind are synonymous terms. One could say that the mind is merely a functioning of the brain.

Brain size is not a direct measure of the mental capacity of animals. The initial structure and the degree of development of the brain are far more important. There are a number of animals in existence today, such as the whale and the elephant, that have larger brains than man. However, a great portion of their brains is occupied with the job of regulating the functions of their large bodies. As a result, less brain area is left over for the higher functions of thought. Brain size does, to a degree, give us some idea of mental capacities. For comparative purposes, the weight of the brain is divided by one ten thousandth the weight of the body. By doing this, the approximate results are as follows:

Fishes 2   Reptiles 8   Birds 42
Mammals 54   Man 280

In addition to brain size, the intellectual capacity of an animal is determined by the number of convolutions in the gray matter of the brain. By forming convolutions, the brain increases its surface area and its ability to store knowledge. Birds and the lowest forms of mammals have almost smooth brains. Dogs, apes, dolphins and elephants have a moderate number of convolutions. Bees and ants have highly convoluted gray matter. In man, a manual laborer will have fewer convolutions that may be half as deep as a highly educated individual. A newly born baby has no visible convolutions, and during youth, those present are imperfect. As the convolutions develop, the brain becomes denser and increases in weight. It reaches a maximum development in man between the age of 40 and 50. As the brain grows older, it begins to shrivel. Gaps form between convolutions. Theses gaps fill with water, and the blood supply to the brain decreases. The color of the gray matter deepens, and its weight begins to fall as brain cells die and are carried away.

The normal fully developed brain of a man weighs about three pounds. The brains of the mentally retarded weigh between one and two pounds. The average European cranial capacity is 90 cubic inches. The Hottentot natives have a capacity of about 65 cubic inches. Some Hindus have capacities as low as 45 cubic inches. This approaches the capacity of gorillas. The

Egyptians, despite their great intelligence, had small cranial capacities. Dinosaurs had small brains in comparison to the reptiles of today.

The existence of a brain is not a prerequisite to the existence of life. Experiments have been conducted on chickens and other animals whereby the brain was removed layer by layer. With each removal, loss of mental capacity of the animal was noted. With total removal of the brain, the animal became little more than a vegetable. It stood motionless in one spot. Yet with artificial feeding, it continued to live for months and years, increasing in size and weight. But remember that without a brain there is no knowledge of existence. From this description, it seems obvious that life, in itself, does not require a brain. Many bodily functions are carried on by their DNA. Certainly, plants fall within this category of living things. Also, there is no knowledge of existence among the cells of living organisms. They perform their functions automatically as they have for millions of years.

The human brain can exist without a supply of oxygen for about six minutes. Some of its parts can survive slightly longer. If the heart function that supplies that oxygen is restored after that time, the brain cells would have already been destroyed, and the individual will be little more than a vegetable. When human heart functions have stopped for short periods of time, those persons claim that they have seen angels and other religious phenomena. They make the assumption that it was Heaven that they were experiencing. Such experiences have been used as examples to prove the existence of God. However, the idea that life ceases with the loss of breath is wrong. As we have seen, the brain remains alive for a short time thereafter. If it is alive, it can still have thoughts. In fact, it may be possible for blood to flow in smaller quantities to the brain by convection or other means. That way, the time that the brain stays alive could be greatly extended. Also, it is only natural that when one is dying, the brain would concentrate on Heaven. Also, remember that many other body cells, such as hair cells, continue growing after death, as long as they receive nourishment.

Long before the first amphibian climbed out of the water onto the land (about 280 million years ago), evolution had produced two eyes, two ears, two nostrils and one mouth in creatures. As simple cells developed into more complex creatures, many combinations must have been tried. The present arrangement must have proved the best for survival. To quicken the ability to survive, the brain became located close to the major sensory organs, but that is about as far as practicality went, for as the animal brain developed over those many millions of years, various functions took up residency in often distant parts of the brain. The right eye is communicated to cell structure on the left side and rear of the brain. The left eye connects to the right rear side.

The mind needs past experience to function. It can be either our own past experience or that which is embedded in the DNA of our ancestors. The ability, or lack thereof, to have religious beliefs is not all taught. Genetics play an important part in the process, just as they do in all aspects of life.

Most people believe that everything that enters their brains is true and real. They don't question where the inputs are coming from. Is it something that one can see, or is it only an idea? In fact, we will see in the following pages that the reality of inputs is open to question.

Development of the brain starts in the fetus, but it is only an approximation of what it will be after its 100 billion or so neurons and 900 billion or so glia cells mature. The sense organs are the first to connect to the processing centers of the brain. By the time of birth, a baby has almost all of the brain cells that an adult will ever have, but at this time, the baby's brain is only about one quarter its adult size.

When a child is born, its brain weighs approximately one pound. Some new brain cells can continue to develop in the brain until the child is between one and two years old. After this time, with a few exceptions, no new neurons are produced. Neurons that are destroyed in later life are not usually replaced. The brain and its cells continue to increase in size until a child is about fifteen years old. The fully developed brain weighs about three pounds.

Again, it is not only the inputs from our sensory organs that affect the development of a child; it is also the genes received from its parents.

Some believe that a soul reincarnates from Heaven into a baby at the time of birth. That same soul has gone to Heaven after death and can remember its life on Earth and its deceased relatives that it meets in heaven; yet it cannot remember its past lives when the soul reincarnates into the new baby. That is inconsistent with atheist reasoning. Also, how can the soul meet its dead relatives in Heaven if they've already reincarnated? Rather, the atheist believes that a child is conceived when a man's sperm and a woman's egg comes together in the womb. The genes presented in the sperm and egg, not a soul, dictate how the fetus shall develop. In the process of growth, the fetus starts with simple cells. These cells divide and divide again trillions of times by the instructions embedded in the DNA. The fetus climbs up the ladder of evolution. Scientists call this ontogeny recapitulates phylogeny. At one point in its development, gill slits similar to those of fish are present. Later, these slits develop into other parts of the body, such as the ears.

The human brain does not record thoughts in a photographic manner. Rather, it compares inputs to previously obtained inputs. The cortex of a newborn baby is nearly blank. Day after day, it receives inputs and establishes patterns of behavior. As time goes on, there are fewer possibilities for the brain to establish alternate ways of action. As one grows older, the randomness of the brain becomes depleted, and it is less able to experiment. In other words, most conditioned behavior patterns are established in the brain during childhood. These are modified very gradually by changes in the environment. Older people find it extremely difficult to develop new conditioned responses. That is why it is important for the church to implant religious doctrines in children at an early age.

When thought processes are examined in more detail, it becomes evident that actually pain and pleasure exist only in our brains. When inputs to the brain are in harmony, we experience pleasure. Pain is felt when a train of nerve impulses arrive at the brain in a disorderly manner. Pain and pleasure are a human interpretation of electrical impulses as they are related one to the other. When these impulses to the brain are out of order, the brain reacts in such a way as to restore the body back to a steady state.

It is not my purpose to enter into a detailed presentation on how the brain works. Rather, I am only interested in the general concept to better understand how it establishes realities. Also, keep in mind that although tremendous strides have been made in understanding how the brain works, we still have a long way to go.

Religious beliefs are developed and stored in the brain. The genes that we receive from our parents have a decided influence upon religious beliefs. If one has some understanding of how the human brain works, he can better understand how man has acquired religious thoughts. Until recent times, very little research was conducted on its functioning. Consequently, today's knowledge is sketchy. However, from what scientists have learned, it is possible to achieve a rough concept of its workings.

As a starter, let us consider the receptors. They are located throughout the various parts of the body. There are thousands of receptors underneath the skin and millions more in the sensory organs, such as the eyes. From the receptors, impulses or small electrical charges are sent along thin nerve fibers to the brain. The nerve fibers are in the form of very long threads, each thread less than a thousandth of an inch across and grouped into a bundle with other threads. These threads are connected to input neurons. Standing guard at each neuron's synapse are astrocyte glia cells which are thought to store memory. Also, astrocyte glia cells are thought to issue commands to motor neurons which fire small electrical charges to activate muscles, etc. We will get into how all the various kinds of glia cells function later.

There is one impulse nerve fiber for every receptor. Each muscle is controlled by hundreds of output nerve fibers, each controlling a small portion of that muscle. The nerve fibers can either respond or not respond. Variations in activity that a muscle exhibits are a function of the number of nerve fibers that have activated it. To make a muscle function completely, a great majority of the nerve fibers must simultaneously send impulses.

The brain sorts out the information coming to it. Most of the information falling on our retinas, eardrums, nostrils, etc. is not processed. The brain picks out only the inputs that it has cell structure in place to process. If the individual is religious, religious inputs will be processed; nonreligious ones will be ignored.

The world we see is an invention of the brain. It analyzes and interprets the flood of information coming into it. Even though our visual inputs appear to be fairly accurate, remember that the picture that our brain paints is just that, an interpretation of reality, and not reality itself. Our brain extracts and stores the constant invariant features of objects. It then compares and interprets the flood of information coming from the retina, etc. to it. In this manner, it constructs a visual world. There is a complex division within the cortical areas dealing with sight. Each division and subdivision deals with a particular aspect. For example, area V5 is specialized to deal with visual motion, and V4 is specialized to deal with color and line orientation. At this time, the complexity of the process is too staggering to be fully understood. However, researchers have been able to discover brain functions because various parts of the brain have been damaged in different patients. From them it is possible to see what functions of the brain have been lost, and thereby determine what functions each part had.

Brain cells require oxygen to function. The oxygen is carried to the cells by the blood. Most of the blood pumped by the heart goes to the brain. Depending on which task is being performed, the blood will shift in the brain to supply the working cells with oxygen. Also, for brain cells to function, various chemicals, such as potassium and sodium ions, enzymes, proteins and amino acids, are required and supplied by the blood. Neurotransmitters, such as serotonin, carry messages between cells. Calcium is also very important for brain functions.

Seeing and awareness is a constructive process. Many parts of the brain carry out complex computations. It produces a symbolic representation of the visual world. For example, you contain a representation of the Statue of Liberty in your brain, which is visually inactive. But if you think about the statue, its cells become active and fire away. An object is represented in the brain in many ways, such as visual images, written words, related sounds, smell or touch. Some cells can be activated more by a face turning in one direction than in another. Seeing a familiar face, a person's sex and facial expressions can be correlated with brain cells firing in various areas of the brain.

Some cells may not fire because they do not have time to react, but the majority can still send an action potential (electrical spike) from one part of the brain to another part of the brain. These action potentials measure about 100 millivolts and can be generated at about 200 per second. It should be obvious by now that a nonmaterial spirit can't in any way influence the operation of the brain. At the same time, it also becomes obvious that gods, spirits and the hereafter are ideas only that are stored in the brain by material brain structure.

By computer standards, brain cells act very slowly. This is partly compensated for by firing many cells simultaneously and in parallel in a hierarchical order. The brain also has the ability of filling in information. An example would be filling in the blind spot in each eye caused by the lack of photoreceptors where the optic nerve exits the front of the retina in people.

Again, each part of the brain has a separate function. The brain parts are interconnected by fibers along which impulses travel. The medulla oblongata is located above the spinal cord and regulates such body functions as breathing and circulation. The slightest damage to its cells causes death within minutes. The cerebellum accounts for 85% of the brain's weight. It receives inputs from the muscles, joints, skin, eyes, and ears and regulates posture, balance, and movement. The cerebellum regulates those functions, which account for man's basic intelligence. The thalamus is primarily a distributing center that relays sensory impulses. Both the thalamus and the hypothalamus lie near the center of the brain. The hypothalamus regulates many basic body functions, which include body temperature, appetite, thirst, and sleep. Damage to the appropriate cells can cause a voracious appetite or sleeping sickness.

At the top of the brain is the cerebral cortex, where most of the brain cells are located. The more complicated information, such as that received from the eyes, ears and receptor organs under the skin, are sorted out at the thalamus in the center of the brain and transmitted to the appropriate cells of the cerebral cortex. Most of the blood pumped from the heart goes to the brain. Almost all of the blood is supplied to the gray matter of the cerebral cortex, which is deposited in five to seven distinct layers.

The frontal lobes, which are located behind the forehead, occupy a great portion of the cerebral cortex. They are much larger in man than in other animals, controlling the finer functions of the brain, such as high levels of communication. Their removal does not impair the basic body functions. If they are removed, it can cause an individual to be more docile, frank, and even rude in behavior. Keep in mind that if the human brain were constructed differently than it is, there is a good possibility that there would be no place within it for storing and processing spiritual ideas. Much controversy has resulted from the use of the operation termed a prefrontal lobotomy, which has been performed to quiet dangerous criminals. If too many cells are removed from the frontal lobes, religious feelings can be completely destroyed. Animals of a lower order appear not to have religious feelings. It may be because they lack a sufficient number of cells in the frontal lobe area.

Much of the activity in the brain is associated with electrical charges. Clinically, small electrical charges can be applied to the surface of the brain causing paralysis, loss of speech or a need to cry out. Loss of thought or hearing of sounds, such as ringing and buzzing, can also be induced by this method. During these experiments, the individual has absolutely no control over his reactions. Many believe that this clearly shows that man does not have a free will. He is controlled by the impulses to his brain fitting memory processes in it and establishing behavior patterns. Man is free only within the limitations of his brain. His actions are determined by impulses reacting with the present memory processes of the brain. If he is programmed to think in one way, he must react accordingly. Of course, the programs or memory processes are continually changing in relation to impulses received. This continual change makes us think that we have free will.

Many of the things that we see and experience are not real. For one thing, colors do not exist outside our brains. They are our brain's interpretation of various wavelengths of light. Many other animals do not see colors. These wavelengths change with the intensity of light; yet our brains have the ability to interpret a constant color for different intensities. The image of objects varies with distance. Yet our brains can visualize their actual sizes. The visual world that the brain constructs is, in many ways, far from real. Exactly how unreal our world is is debatable.

Also, there is no such thing as sound in the real world. Sound is produced in our brains in response to varying pressures on our eardrums. It is obvious why hearing would become an evolutionary development in animals. Those who could not hear their enemies approaching would soon become eliminated.

The brain can play tricks on us. We may think that we have seen one thing, but when we look closer, it turns out to be something entirely different. In fact, the brain can manufacture whole pictures that do not exist outside of it.

A neuron brain cell collects signals from other neurons through fine inputting structures called dendrites. There may be many dendrites reaching a cell. When activated, the neuron sends out a spike of electrical activity along a long, thin strand called an axon. The axon can split into thousands of branches, which are called synapses. Synapses are held close but not in contact with the dendrites of other cells by glia cells (astrocytes). Learning occurs by the changing of synapses so that the influence of one cell on another changes.

Again, the basic process is that neurons receive input signals into their cell bodies through their dendrites. When the neuron cells reach an action potential, they discharge a signal out along their axons. There may be many dendrites entering a cell, but there is only one axon exiting it. The axon then branches out to the dendrites of other neurons. These connections between neurons change with new inputs to the brain. Also, axons (synapses) move to new neurons if neurons die.

There is no screen upon which the brain's final processed information is projected, but rather, the various layers connect directly and reciprocally with each other, forming a model in the brain that represents human reality. Of course not all areas of the brain communicate with all other areas of the brain. That is why it is possible to have, at the same time, religious thoughts and scientific thoughts that would normally conflict with one another and still have no mental conflict.

The use of hallucinogens, such as LSD (lysergic acid diethylamide), alter the functioning of the brain. By using them, it is possible to see brilliant colors, wave-like patterns in the atmosphere and many more abnormal happenings. Nature, in general, appears to fit into a tightly meshed pattern. Some believe that these happenings are closer to reality. It is more likely that they are only another interpretation of matter that is even further from reality than the mental processes that would normally occur.

[SEE THE PARTS OF THE BRAIN ILLUSTRAITON]

[SEE BRAIN CELLS ILLUSTRATION]

The center of the retina of each eye has nearly half a million light sensitive cells. These cells connect to nerve fibers, along which impulses pass to the brain. An increase in the amount of light falling on a cell increases the number of impulses passing to the brain.

The most important inputs to the brain come through the eyes. Through the input of images, the brain invents a visual world. A tree at a distance looks far different than one close up. Our mind knows what the leaves look like, so it can fill in that information, even if it cannot see them correctly at a distance.

The retina is mostly connected to one area of the brain called V1, which contains a map of the entire retinal field. From there, connections are made to six layers of cells. At birth, V1 is fairly well developed, but the other layers depend on the acquisition of experience to develop.

The cells of V1 and V2 assemble and pigeonhole the signals coming from the retina and then send them to their appropriate layers. These layers then send signals back to V1 and V2, as well as other parts of the brain.

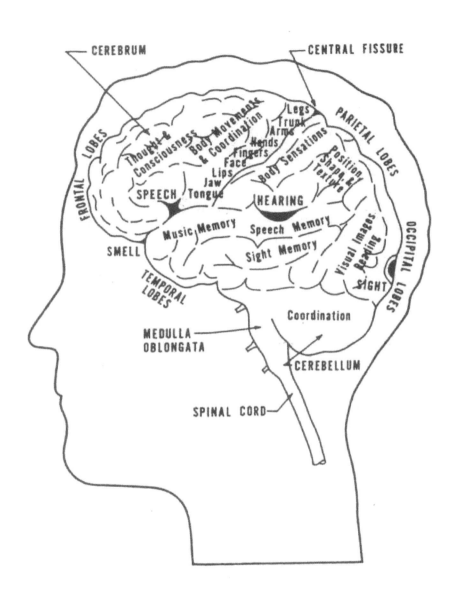

THE PARTS OF THE BRAIN

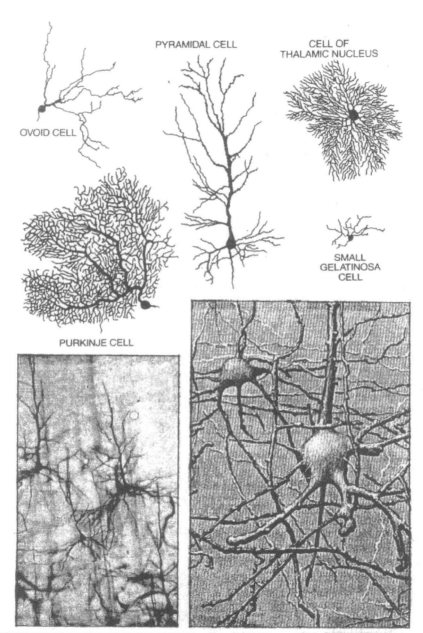

OVOID CELL

PYRAMIDAL CELL

CELL OF
THALAMIC NUCLEUS

SMALL
GELATINOSA
CELL

PURKINJE CELL

NEURONS

Reprinted from SCIENTIFIC AMERICAN
September 1992, Vol 267, Number 3,
415 Madison Ave, New York, NY 10017

We may see someone's face and match that vision to information in our brain. "Hello, George." George responds, "I am Bill. My twin brother, George, is over there." We then look at the two brothers and discover that Bill has a mole on his face. We have now learned enough to distinguish the two. There are many other differences between the two, but our brains stops comparing because we have learned enough to achieve our purpose.

What happens in the brain is not like a computer, and what we see is not like a TV screen. What has happened is that the various functions performed by the brain have developed over many millions of years in different areas of the brain. These functions interrelate, thus producing the whole picture. Such things as movement, color and shape are processed in separate parts of the brain. Other brain parts concern themselves only with one thing, such as the recognition of faces. The information stored by the brain is fragmental. The amount of information stored in it depends on the learning process. Here, again, when we think of the fragmental way in which our brain works, we wonder how real its thoughts are.

The area of the human brain that is related to language is in three parts. One deals with non-language concepts, one deals with the construction of words and sentences, and the third integrates the first two. Although other animals can create sounds, they lack the human complexity to produce speech. Without this speech complexity, it is dubious whether any of the other animals have developed an idea of gods, spirits and a hereafter.

It takes a much smaller number of brain cells to form words and sentences than it does to form concepts. To some degree, this may be due to the late development of language in the evolution of man. The area of the brain that generates speech and written text is largely located in the left hemisphere. Damage to this area can cause a loss of the ability to form words and even the ability to use sign language. Again, by observing which damaged areas cause what loss of ability, we can determine what areas have what function. Even a simple function of the body, like raising a cup of coffee to the lips, is generated by neurons firing in many parts of the brain.

The construction of bird and reptile brains differs from those of humans, but the basic principles are very similar. Evolution has built parallel computing into almost all animal brains. Parallel computing is receiving signals into the brain from all of the senses at the same time. A good example of parallel computing involves our eyes. Each light receptor works in parallel with the others, sending parallel signals all at the same time, which are processed into the elaborate images in our brains. Because of the rapid inflow of images, our brains can perceive them as being out beyond our eyes instead of inside our brains.

Now we can record the activity of a single neuron or group of neurons. We can relate neural activity to a specific mental state, such as seeing a face or color or line. All this has been made possible with PET scans and FMR scans. It can be determined how molecules in microscopic neuron circuits perform doing mental tasks. Researchers can identify the genes necessary to produce and deploy those molecules.

As I have stated, we do not perceive reality; we interpret it. How different brains interpret inputs depends on how they are constructed. Different animals receive different sensory inputs. Some animals receive sets of inputs that other animals do not. This creates a different world for them. Many insects have compound eyes sending many signals to their brains. Our eyes have rods to distinguish the level of light and the forms of what we see and cones to measure color. Also, if an animal combines different sensory inputs in its brain, it might get a different interpretation than if it separated them. For example, heat and light might be combined, and they would be sensed the same. One wonders if separating inputs in the brain is closer to reality or not.

Just alone by understanding how brains work is enough to know that we are not getting a real picture of the world that we live in. We are only interpreting it.

We have different kinds of memory. We have short-term or temporary memory, intermediate memory and long-term memory. We have memory of facts and memory of events, and there are other types of memory also.

Establishing long-term memory requires that certain processes take place at the molecular, neural and glial levels. Also, specific genes must synthesize fresh proteins to consolidate the memories. The formation of new memories involves the formation of new synapses connecting different neurons and the monitoring of those synapses by different glia cells (astrocytes). The cells are highly parallel and deal only with particular classes of events, such as a head turning to the left. There doesn't seem to be a central area of the brain where all activity comes together. The closest might be the thalamus area of the lower brain. However, that area does not receive inputs from all other areas of the brain.

Our lower brain (thalamus, hypothalamus, etc.) deals mostly with drives and feelings. Our cerebellum deals mostly with physical movements. Our neocortex is concerned with combining and understanding the inputs from our senses.

Most neuroscientists have a problem understanding consciousness. They can't explain it. If astrocyte glia cells can store memory with their calcium puffs and calcium waves, then they are most likely also responsible for consciousness. It is also known that the glial cells in the hippocampus can store long-term memory for weeks, but it gradually transfers memory to other areas of the brain.

The brain contains various kinds of memories. Probably the most important to human mental development is called the working memory. It is located in the prefrontal lobes of the cerebral cortex. Its cells store the information necessary to perform mathematical computations and other higher-level mental activities. These cells are capable of building abstract concepts, such as gods, spirits and the hereafter, even though real objects are not present. The prefrontal cortex mediates between other parts of the brain, including those where long-term memory, associated memory and learned memory are found.

Memory cells have the ability to hold the visual space coordinates of objects, even after they have vanished. The cells of the working memory appear to be located together in one area of the prefrontal cortex. Each cell deals with one specific aspect. The working memory also excites or inhibits activity in other parts of the brain, such as the motor centers that cause movements of the body. That is, an object's location, or its features, such as its color, size and shape. The prefrontal cortex is probably divided into many areas, each dealing with a different kind of information.

Neurons appear to have no other function than to relay signals from one part of the brain to another part of the brain. It is also apparent that most astrocyte glia cells have memory capability. That capability may be short-term, medium-term or long-term memory. Some astrocyte glia cells are capable of transferring memory to other parts of the brain.

Once the brain has been programmed with inputs from the outside world, it can develop thoughts and even whole concepts that exist inside the brain and not outside of it. The brain can develop unreal thoughts from real inputs, but it can not develop unreal thoughts if there is no information in the brain to base them upon. For example, one could not develop ideas about UFOs if one had never heard about them or seen objects that resembled them.

Even without any damage to the brain, the thoughts produced in it are often far from real. Unreal ideas come into our brains mostly by sight and sound. Once stored there, they are often interpreted as being real.

Brain cells are the only form of matter that have any real knowledge of their existence. Also, it takes many billions of brain cells programmed in a certain arrangement to accomplish that feat. Again, spirits have no material composition and cannot enter the brain. Only material inputs can do that.

Brain cells cannot survive without a flow of blood to them. The blood carries oxygen and an array of different minerals and chemicals to nourish the cells. In other words, the brain is a material entity. It functions by material inputs coming to it and being processed by it. Again, when a brain dies, all knowledge that it stored in it is lost. It is like a computer that goes down. Everything that is stored in the computer is lost.

Because of the way the brain functions, it is possible to see things inside our brains that do not exist outside of it. This is especially true of dreams. They are based upon the information stored in our brains. Some people have such vivid dreams that they believe their dreams are real, rather than a construction in their brains.

Now that the basic functioning of the brain has been discussed, one is better able to examine some of the facets of thought. It becomes easier to see how people can be influenced in life. It is also easier to understand why people believe what they do. For most individuals, their brains have reached a state of certainty before they have passed their teens. In fact, most thought patterns have been established before the age of ten. Every new idea encountered in later life must fight an uphill battle for acceptance. In old age, the ability to conceive whole new concepts becomes next to impossible.

In our society, a normal individual lives within a limited scope of activities. This scope is shaped by the background, by the education, and by the environment of the individual. If this scope changes, it usually changes very slowly. One is considered normal by the standards of society if he or she accepts most of the behavior patterns and social standards of the community. In other words, a normal person is susceptible to suggestion. One who holds minority opinions or acts contrary to society is considered eccentric or mad. Such individuals are difficult to influence by suggestion. During past centuries, many atheists were included in this category of individuals difficult to influence by suggestion.

Since the normal brain has the capacity to adapt to an ever-changing environment, it is actually more susceptible to thought control than an abnormal one. Many persons try to influence us through the use of thought control. Advertising confronts us from every angle. We are forced to conform to the laws of society. It is sometimes difficult to see how thought operates. The mechanics of man's thinking can be better understood by observing the techniques of brainwashing.

To be successful at brainwashing, the brain must be brought to the point of exhaustion. When the brain reaches this point, it becomes incapable, for a time, of intelligent reasoning. The normal positive and negative conditioned responses are represented in the brain by cell structure. Under normal conditions, the various parts of the brain work only part-time with periods off for rebuilding themselves. When these cells are subjected to exhaustive use, they are destroyed or their material composition is altered so that they can no longer function as in the past. It is then that the physical control is relinquished and exhaustion results. If a sudden collapse can be produced after a prolonged or intense emotional stress has been imposed, a great portion of the brain can be wiped clean of any recently implanted behavior patterns, and new patterns can be developed in their place. Former positive conditioned responses can become negative and vice versa. Political brain washing, such as the Communists used, works this way. Through fear, anger and other emotions, high stresses are built up in the brain. This temporarily impairs judgment and heightens suggestibility. To make the conversion, there must always be an escape route given for the mind to follow. Such a route is the promise of the Communist utopia. To avoid backsliding, severe penalties must be fixed in the mind.

Police have cross-examined suspects for long periods of time. By using threats and disallowing sleep, the suspects were brought to the point of collapse. In order to find an escape route, the suspects signed confessions for crimes of which they were innocent. Now with the advent of DNA testing this technique is becoming more and more obsolete.

In religion, the most successful preachers are those who can arouse a feeling of guilt among their congregation. The preaching of hellfire and damnation produces a mental stress that must be and is relieved by the promise of eternal salvation through faith. To prevent backsliding, severe penalties are fixed for sinning. On the other hand, intellectual religious indoctrination without emotional excitement gains few converts.

People who are tense and anxious seldom check religious doctrines for consistency of details. They accept most everything told them without question. Atheists, on the contrary, usually fit every so-called truth into logical formulas to arrive at their answers. If inputs do not appear to fit in, he or she rejects them without hesitation. If the atheist's mind has been programmed to believe that the teachings of the Bible are nothing more than fairy tales, then most everything associated with religion might be rejected.

As mentioned earlier, the basic thought process seems simple; it is the billions of brain cells involved that make it complex. Thought processes are determined by the impulses that are received in the brain. If something exists for which the brain cannot receive impulses, knowledge of that thing is not available. The concept of the existence of God could have been passed along from antiquity. If only the concept of God is received by our senses, then God, Himself, is not being received, and God does not exist.

The two most common illnesses of the brain are schizophrenia and manic depression (mania). Patients of these illnesses commonly hallucinate and exhibit periods of depression. But besides these illnesses, there are other factors that could cause so-called normal people to hallucinate. To hallucinate is to see things inside the brain that don't exist outside of it. Examples would be dinosaurs, aliens, gods and spirits. Stress can play an important part in bringing on hallucinations. People with extraordinary mental genius seem to be more prone to periods of mental illness. Patients with mental illness appear to have a less than normal flow of blood to the prefrontal cerebral cortex. Also, the hippocampus, which connects to it, has larger spaces between its convolutions. This suggests that brain damage has occurred and brain cells have been lost. Dopamine is a chemical in the brain that regulates how brain cells react to stimuli. A lack of dopamine can cause schizophrenia and deficits in working memory.

As one ages, the brain loses cells, and chemical alterations occur. However, unless a disease such as Alzheimer's, Parkinson's, ALS or multiple strokes occur, there need not be any noticeable decline in intelligence. It is normal for brain cells to die and their functions to be replaced by other cells. Unless reproduced by astrocytes, neurons do not reproduce themselves, but then we have plenty of them to start out with. When neurons die, their neighbors grow dendrites to pick up signals coming to the dead cells. The various parts of the brain lose cells at varying rates. Some areas lose only 5% during a ten-year period. Parkinson's disease, on the contrary, can destroy

70% or more. Glia brain cells do reproduce themselves, but as one gets older, it becomes harder for them to do so.

When the human body dies and the heart stops pumping blood to the brain, the brain dies. After approximately 4 to 6 minutes of complete blood starvation, the brain cells cease to function, and all knowledge is lost. The body can be artificially kept alive, but all awareness of human existence is gone.

Again, our brains do not see reality; they interpret it. Based upon the inputs that the brain receives, it is capable of constructing a world that does not exist. An example of this would be out-of-body experiences.

We have no knowledge of the universe before conception because there is no brain to store that knowledge. Any knowledge that we have of happenings before birth is learned. Conversely, after we die, all knowledge we have stored in our brains is lost. There is no longer a living brain to process it. The matter that composes us and the universe still exists, but the inputs from that matter can no longer be processed by the dead brain. When we die, the universe ceases to exist for us!

The atheist believes that gods, spirits and the hereafter, which have been defined to be nonmaterial, do not exist. The atheist believes that the idea of gods, spirits and the hereafter was generated in the human brain and stored by human brain cells to explain the otherwise unexplainable.

Man could shape clay in his image, but he could not breathe life into that clay. He required a god to do that, and that life was a spirit, and the spirit controlled man's body. Today we know that matter composes and controls our brains and bodies. Everyday we are learning more about reality and science, yet many do not accept its discoveries. It is true that scientific knowledge is based upon assumptions, but at the same time, scientific knowledge is the most accurate knowledge we have. One thing that seems sure to the atheist is that we cannot perceive nonmaterial entities. If gods, spirits and the hereafter were real, our instruments would be able to detect them. Maybe we would even find the location of Heaven and Hell in the universe and count how many billions of people were sitting at the right hand of God.

Up until now we have been talking mostly about neurons. Some scientists still think that neurons do it all, but that picture is changing. Sure, neurons number some 100 billion cells, but the other cells collectively called glia cells number over 900 billion cells.

How did they ever come to be called glia cells, since there are over eight very different cells in the group? Well, back in the 1800s when scientists first discovered them, it was thought that their only function was to hold neurons together. So, they called them glia cells, which is Greek for glue.

Now what do glia cells do? Collectively, they support other cells in many ways. They clean up brain debris; they bring nutrients to cells; they digest parts of dead cells; they regulate intercellular space; and they provide insulation (myelin) for neuron axons.

Glia cells have only one process, while neurons have two: axons and dendrites. However, glia cells can engage in a two-way dialogue from the embryonic time until old age. They communicate among themselves in a network parallel to that of neurons. However, they use chemical signals or small electrical ones.

More specifically, "Schwann cells" and "oligodendrocytes" myelinate (insulate) neuron axons.

"Muller cells" I'm not quite sure of. However, they may have something to do with signals coming form the retina of the eye.

"Epithelial cells" prevent blood contaminates from entering the brain.

"Ependymal cells" and "tanycytes" work on ventricles and blood vesicles and form tight blood junctions.

"Microglia", which is the smallest glia cell, responds to infection and injury in the brain.

"Astrocytes" outnumber all other cells in the brain, outnumbering neurons by almost two to one. We will talk about them next.

"Sensory neurons" receive inputs from many parts of the body, including the eyes, the ears, the nose, etc., etc.

"Motor neurons" receive inputs from astrocytes to send commands to the muscles and other parts of the body.

The cell that we will be talking about for the remainder of this chapter is the astrocyte. It is becoming very evident that it is the cell that stores memory and maintains consciousness. The astrocyte is a rather round cell with lots of spidery protrusions that are frequently in contact with blood vessels. Neurons need astrocytes to survive, but not the other way around. Astrocytes furnish neurons with the energy they need from the blood.

As mentioned before, neurons do not reproduce themselves, but glia cells do, and in some cases astrocytes reproduce neurons.

We also know that when a neuron reaches an action potential it is an astrocyte that tells it to fire. But how does the astrocyte store the memory and process it to tell the neurons to fire? It is really not that clear to me, but I will present some of the facts that scientists have discovered.

Astrocytes monitor the information coming to the synapses of the sensory neurons, so they know what is happening in the body. Neurons store calcium at their synapses. Astrocytes also store calcium in their endoplasmic reticulum. The calcium is released, first forming a calcium puff, which then spreads to other astrocytes (maybe 100), forming a calcium wave. Is this where memory and the other functions of thought are stored and processed? I don't' know.

At some point, astrocytes become impregnated with fluid, and glutamate is released from the astrocytes, causing motor neurons to fire, making muscles flex, etc., etc.

Also, it might be noted that alcoholic beverages do not kill neurons to the extent that was originally thought. However, they do kill glia cells. But then, glia cells can reproduce themselves. As mentioned before, the problem with old age is that the glia cells do not reproduce themselves at nearly the rate that they did when we were young.

Glutaminate is not the only transmitter. Depending on the part of the brain, it could be dopamine, serotonin, etc. For an astrocyte to take in glutamate two or three sodium ions and one hydrogen ion must also come in, and one potassium ion must go out. I wont' bore you with any more of this complicated information. If you're interested, there are many books on the subject, which you can locate on the internet or at your library.

Again, it has taken millions and millions of years for anything as complicated as the human brain to evolve. That is a very long time before man got the idea to invent gods.

11. RELIGION AND THE CHURCH

Religion is a system of belief in the existence of a divine or supernatural power or powers. It is a belief in a supreme being or beings to be obeyed and worshipped as the creator(s) and ruler(s) of the universe. Religion often involves a code of ethics, a philosophy and a belief in life after death. Under a religion, some would include such cults as Devil worship. As mentioned earlier, there are probably as many different ideas of what religion is as there are people to have them.

Many believe that behind the universe there is a supreme power that works out the destiny of mankind. If this is correct, then man is merely carrying out a divine will. Can he, under these circumstances, be responsible for his actions? When he murders, rapes or commits other crimes, he is only carrying out this divine will. When he does something praiseworthy, he is not deserving of praise because it is God's will. At one time, the churches preached the doctrine that God was everywhere. He filled space. Nothing was accomplished without Him. If this were so, that God controlled everything, then He also controlled man when man sinned. Man was not responsible for his sins. God was the one who caused them. Also, when man questioned the existence of God, God was questioning his own existence, along with man.

Nowadays, theologians tell us that man is a free agent. He has the power to choose his actions; to do good or bad. He has free will. God may help him, and the Devil may tempt him, but in the end, man must make the choice. But, in reality, isn't free will an illusion? In the last chapter we learned what man knows about the functioning of the mind. We learned that every action is caused by inputs to the brain, fitting and interacting with the memory processes that are established in the brain. In other words, our actions are controlled by these memory processes and the inputs they receive. There is no real free will of action.

As can be seen from the tables on the following pages, most of the world's population adheres to one religion or the other. Year by year, this religious enrollment is growing weaker. It is mostly in the Muslim countries and India where population is on the increase, and religion is accepted almost universally without question, that an increase in religious enrollment is seen. These statistics may be misleading due to several factors. In the USSR and China, many people may be listed as atheists who are actually religious. In other countries, people may belong to more than one religion, thus counting themselves twice. Still, in other countries, atheists and agnostics may count themselves as belonging to a church for business or social reasons. Some countries count all their people as belonging to the state religion. Even if the figures are not entirely accurate, they do give some idea of how religions are moving.

Most people in the world have a religion that was handed down to them from their ancestors. In many cases, they do not truly understand their religion. They follow it through routine and make little or no attempt to examine it. There are others who have a need to explain everything that they experience. If they cannot explain things by their knowledge, they can always use religion's explanation that "it is God's will." Religious people often believe that plagues, famines, wars, floods and so forth, are demonstrations of God's wrath. Even insurance companies call them "acts of God." The religious person prays and makes offerings to appease the angry god. The atheist doesn't waste his time in prayer. He takes what steps are necessary to correct his problems.

By definition, God is incomprehensible to man. If man was not constantly reminded of God, if man did not need God to explain the unexplainable by the expression "God did it," and if man did not need God to insure life after death, then man would not have a need for God.

When one thinks about it, all babies could be considered atheists. At birth they have no concept of God. The mind of a baby is almost blank. The idea of religion is handed down from father to son or church to children and is part of the social order of the community. If one wants to belong to the community, one must belong to the church. Contrary to what some may believe, religious ideas are not inherited; they are taught. The child may inherit a type of brain that has more capacity or less capacity for experiencing religious feelings, but that is about as far as it goes.

In other words, not everyone is born with the same amount of marbles.

Again, the idea of God is not innate in man. Children raised experimentally without any contact with their fellow humans have no idea of God. In most cases, the idea of God is planted in the minds of children at an age when they have little ability to reason for themselves.

It is not surprising to find man still believing in God after so many centuries of scientific change. For one thing, man's basic physical construction and needs have not evolved nearly as rapidly as science. There is also the factor of the law of inertia, or continuance. That is, matter and everything composed of matter tends to continue doing in the future what it is doing at present.

If man finds any conflict between religion and science, he can always use a system that in bookkeeping is called "double entry". One relies on science for a living and on God for security. Of course, one must never justify the entries, and one must consider only one entry at a time, so there will be no confrontations with discrepancies.

It is possible to think that if there is a God and man is made in God's image, then God must be imperfect, since man is imperfect. Certainly all of the supposed deeds of God support the contention that he is imperfect. As already noted, the atheist prefers to believe that the reason God is imperfect is because man made God in man's image.

Many will say that God is intangible and does not consist of matter. How can an immaterial God create a material world and regulate a material universe? Isn't it so that man can only perceive matter acting upon matter? Is it not so that rather than perceiving God, man perceives the concept of God which comes to him by reading or by word of mouth?

Even in antiquity, God did not show himself to a whole nation. Only a few favored persons were allowed to hear his instructions. Was this so, or could it be that these few favored persons pretended to hear God's instructions in order to gain fame and fortune? If God had revealed himself to man, why shouldn't he have given the same commandments and instructions to all the people of the world? A universal God should reveal a universal religion. Instead, one finds the various faiths fighting among themselves because of their differing beliefs.

All religious denominations claim to be founded upon the authority of God. Each one claims to be the true religion. Again, if there is a God, he must be very imperfect to give such conflicting instructions. Or could there still be many gods, as our ancestors thought there were? The different gods could each be giving conflicting instructions. Could it also be that if there is a god, the existence of this God is so foreign to any idea that man could have about Him, that he could not have revealed any truth to man at any time?

There are a growing number of people who do believe in a supreme being or power controlling the universe. They believe that this controlling power is beyond the limited comprehension of man. Such things as UFOs, which man has not adequately explained, have tended to fortify this belief.

One could go on asking questions that were asked by man through the centuries concerning the inequities of God. Some questions may seem primitive now. However, it was the answers that caused many to turn to atheism.

A savage, when asked what makes a phonograph play, might say, "God makes it play." Since his ego demands an explanation, he uses God to explain the unexplainable. In the same way, God is used to explain the universe by more advanced people. God is that intangible, that something beyond comprehension, which has more intelligence and more knowledge than man.

Religions need miracles and incredible happenings to impress the minds of the parishioners. Buddha and some other prophets were not nearly as popular in their own time. They were merely men. It took many hundreds of years to build the fantastic stories that elevated them to the position of gods before they could reach the height of their popularity.

ESTIMATED WORLD RELIGIOUS MEMBERSHIP

1964

RELIGION	MEMBERSHIP	PERCENT
CHRISTIAN	950,550,000	30.0
ROMAN CATHOLIC	584,493,000	18.4
PROTESTANT	224,065,000	7.1
EASTERN ORTHODOX	141,992,000	4.5
MUSLIM	455,785,000	14.3
HINDU	395,191,000	12.4
CONFUCIAN	350,835,000	11.1
BUDDHIST	161,856,000	5.1
SHINTO	67,155,000	2.1
TAOIST	51,305,000	1.6
JEWISH	13,121,000	0.4
NO RELIGION, MINOR & PRIMITIVE RELIGIONS	732,357,000	23.0

1974

RELIGION	MEMBERSHIP	PERCENT
CHRISTIAN	944,065,450	24.0
ROMAN CATHOLIC	532,582,000	13.5
PROTESTANT	322,181,850	8.2
EASTERN ORTHODOX	89,301,600	2.3
MUSLIM	529,108,700	13.4
HINDU	514,432,400	13.1
CONFUCIAN	205,976,700	5.2
BUDDHIST	248,516,800	6.3
SHINTO	62,149,000	1.6
TAOIST	31,388,700	0.8
JEWISH	14,386,540	0.4
NO RELIGION, MINOR & PRIMITIVE RELIGIONS	1,390,224,710	35.2

Reprinted with permission from the 1965 & 1975 BRITANNICA BOOK OF THE YEAR, copyright 1965 & 1975 by Encyclopaedia Britannica, Inc., Chicago, Illinois.

2002

RELIGION	MEMBERSHIP	PERCENT
CHRISTIAN	2,014,000,000	33.9
ROMAN CATHOLIC	1,100,000,000	18.5
PROTESTANT	330,000,000	5.5
EASTERN ORTHODOX	220,000,000	3.7
OTHER	364,000,000	6.2
MUSLIM	1,100,000,000	18.5
HINDU	990,000,000	16.6
CHINESE	225,000,000	3.8
BUDDHIST	360,000,000	6.0
SHINTO	52,826,000	1.0
JEWISH	17,263,000	0.3
NO RELIGION, MINOR & PRIMITIVE RELIGIONS	1,188,250,000	19.9

THE INTERNET

Let us take a break and reflect on population growth. In the past ten years our world population has grown by almost 1 billion people to a total of about 7 billion. If the present rate of population growth were to continue for the next 100 years our total world population would reach more than 25 billion. It is questionable whether the world resources could support that many people. It is the author's opinion that we are approaching the time when we must take a hard look at birth control.

Back to the church. Do miracles actually happen? Science has made miracles difficult to believe. If miracles do not happen today, then the miracles that were supposed to have happened as told by the Bible probably did not happen either.

The realm of the church is based upon faith and faith only. To have faith in God is to have faith in someone or something that is not seen or heard. The authority of belief rests in the divine revelation of the Bible, a book that is far from being infallible. The nonbeliever finds it difficult to believe in something he cannot experience and the proof of which is imperfect.

It is not surprising that religious people today do not have the faith that they had centuries ago. Now they want to stay in this miserable world as long as possible. They may pray to God to be healed, but in most cases, they will go to a doctor first. It is this world that is the only world of which they are sure.

Again, the average person does not love God as much as he fears Him. Man is afraid that if he disobeys God's laws, he will go to Hell. Man is God's prisoner, obliged to devote a great portion of his life obeying God's ordinances. However, man seldom thinks of God when he is in the process of sinning. If he thinks of anything, it is whether he can get away with his crime as far as the laws of the land are concerned. Man fears more what he sees than what he does not see. Then, too, one can repent his sins and still go to Heaven, if there is such a place. And if that doesn't work, the church will guaranty a place in heaven for a portion of one's money.

As stated earlier, religion is most successful when fear and tension is instilled in the parishioners. A craving for religion, or a spiritual hunger as it is often called, may be accompanied by insecurity within the individual. This insecurity may be bred by the church itself, or it may be caused by any number of factors in the individual's life. In any event, the church offers a release from the resulting tensions through the promise of eternal salvation.

In the middle ages, religion was used as a tool by the wealthy to control the peasants. They were told that the more they suffered on Earth, the better they would be in Heaven. Their eyes were fixed on the skies to prevent them from seeing the real causes of their suffering. They were told that they were not made to be happy in this world. Life is but a passing thing. They were told that the Earth was not their true country. Their true place was in Heaven. They were told that their sovereigns obtained authority from God; consequently, they were not responsible to anyone but God. For that reason, the peasants should not resist their sovereigns.

Matthew V:28 contends that mental adultery is as sinful as physical adultery. Many religious sects have applied this rule to all of the commandments. Since it is nearly impossible not to have sinful thoughts, the faithful were kept in a continuous state of tension and made dependent upon their spiritual advisors to relieve their feelings of guilt.

The people of Europe were told by the church that the black plague was a punishment for general wickedness. If they did not repent their sins and become good Catholics, the plague would return. In reality, the people needed better sanitation and less superstition. They were also told that schizophrenia was caused by possession of the Devil, or God's vengeance for sinful acts. During the middle ages, men believed that sins were of the body and not of the soul. Consequently, they tortured their own bodies in retaliation for these sins. On the contrary, it is most likely that their sins existed in their minds.

It might be asked, isn't praying to God asking him to change God's plan for the universe, which man does not find beneficial to his interest, in man's favor? The nonbeliever does not believe that there is a God to hear and answer prayers. To him, the universe is governed by laws of nature that are not altered in answer to prayer. If a prayer is realized, it is because someone or some natural force caused it to be realized.

The nonbeliever believes that only matter can act upon matter, and only matter in so-called living form can exhibit intelligence. Is it possible to make the assumption that a soul or spirit without matter can exhibit one's intelligence? Can this nonmaterial entity leave one's body and continue on for eternity? Of course, since a soul has no matter and does not occupy space, there is no problem of finding a place to put all those billions of souls in the universe. However, to the author, the idea of a soul does not sound realistic. Certainly, the idea of personal continuance is not supported by scientific knowledge. Such a belief must be based only on faith. In actuality, when the lights go out that is the end of life.

It is commonly held by a number of religions that animals other than man have a brain but not a soul. It is true that other animals lack the large frontal lobe area of man's brain that controls his finer thought processes and morals. On the other hand, in many ways, other animals experience the same feelings that man does. It seems rather egotistical to claim that man has that special thing without substance called a soul and that his brother animals do not. Also, only living things with brains know that they exist. Plants, for example, grow by the actions of genes (DNA). The DNA is wrapped inside the nucleus of each cell and responds to changes in the environment. The base pairs inside the DNA does not have brains either.

The average person does not want to die. Man's sense of self-preservation is very strong. It is sometimes referred to as the "first law of nature" or in science as the "law of inertia". Objects tend to continue doing in the future what they are doing at present, that is, to follow the path of least resistance. It takes an opposing force to move them from their normal path or to cause death. It might very well be asked, is the fear of ceasing to exist more distressing than the thought of not always having existed? Of course, the answer is yes, it is. The law of inertia has no hindsight, and man's memory only extends back to childhood. He has no knowledge of or a desire to know where all the substance that composes him came from or where it was 5 billion years ago. To many, death is a wondrous gain. While it might rob one of consciousness, it also does away with the pains and sorrows of life.

The Buddhist goal is to reach a state of nirvana, or complete mental nonexistence. Again, it is difficult for an atheist to think of billions of souls floating around somewhere in space, continuing their lives in a better way.

Heaven has no guarantee, other than that found in the imagination of man. Man wants to continue life in another world that has a more durable state of happiness. If God could not remove evil from this world, what assurance is there that He has done so in the next?

Nevertheless, religions of the world remain extremely important to the people of the world. They give guidance and establish a meaning of life. Both Buddhism and Christianity represented the beginning of a new period of change in the world; a change from sacrifices and bloody massacres to the principles of "love thy neighbor." Christianity was devised as the universal religion that everyone could accept. To a great degree, it fell short of achieving those goals. Both the Jews and Arabs rejected it. On the other hand, Christianity became a major force in establishing a code of ethics in the world.

Just before the year 2000 elections, a survey was conducted to determine just how many people believed in God. The survey came up with the figure of 95%. After hearing that, the candidates were quick to express their belief in God. That figure may not be accurate, since going to church in many places is more of a business and social function. Also, other areas have shown that church attendance has fallen off drastically. However, it is clear that the majority of Americans believe in some form of God and hope for a hereafter.

12. MORALS IN THE WORLD TODAY

In the eye of the nonbeliever, the goal in today's world should be to achieve relative happiness for himself and his fellow creatures. Happiness can take many forms. It could be achieved by hitting oneself over the head with a hammer because it feels so good when one stops. Of course, that is extreme and not practical. Happiness could be achieved by working extremely hard, or it could be achieved by sitting in one place doing absolutely nothing. There are as many different ideas of what happiness is as there are creatures to have them. The objective of this life is to be happy in this world for, as the nonbeliever thinks, there most likely is no other world. Happiness, as the author sees it, is usually gained through achievement. Many of the ideas presented in this chapter are those of the author and not necessarily those of other atheists.

The following ideas could be used to replace many of the rules of religions. As previously stated the laws of the land go far beyond the rules of religion, and if the rules of religion were to disappear, nothing would be lost.

The author feels that everyone should be allowed to think in the way that he wants, without harassment. Remember, no matter how ridiculous a philosophy may seem, there is always the chance that it has some validity. Along with this, everyone should be allowed to act in whatsoever way they wish, as long as their actions do not interfere with the freedoms of others.

A need for morals arises from a conflict of desires. Law reflects community morals. Law reconciles the conflict between individual desires or morals. Many laws grew from the idea that the acts they prohibited were displeasing to the gods and consequently might bring divine wrath upon the community. Today, man's scientific outlook makes many laws seem absurd. These laws are still on the books, but they are seldom enforced.

Morals are dictated more by fear of ridicule than by a fear of damnation in Hell. The church may tell us that without God there can be no moral obligation. However, moral obligation comes from the relationship of people associating with each other. Man's morals grow out of the necessity to protect his welfare.

They stem from his love of pleasure and fear of pain. Being religious doesn't necessarily improve morals. Is it not so that a religious sinner can often confess his sins and be cleansed of them or even make a settlement with Heaven by leaving a portion of his fortune to the church?

An atheist does what he thinks is right to please himself and his fellow man. If he sins, it will bring upon him the wrath of others and the punishment that is prescribed by the laws of the state. Morality is acquired through the process of growing up and learning within a society. One's environment determines one's morality.

In recent years, the morals of the world have been undergoing change at an accelerated rate. These changes have been under fire from many segments of the population. Taken subject by subject, the protests may or may not be coming from a majority of the people. However, taking the protests as a whole, it becomes evident that the majority in the world today is not in favor of personal liberty. They want to force others to comply with their beliefs. Sooner or later it will become evident that if individuals are to gain personal freedom for themselves, it will be necessary for them to grant the same freedom to others. It is within the keeping of this objective that the subject of morals is presented.

The author believes that morals should be grouped in different categories. Some morals are, by nature, worldwide. Others are regional, state and local. Still others are strictly personal morals. The author believes that many morals now controlled by society, as a whole, should fall into the category of personal morals.

For millions of years, man and his ancestors have associated with their kinds in little more than a sexual way. It is only recently that man's cerebral cortex reached its present level of development, whereby speech and other higher brain functions became possible. From an early age, children are taught the moralities of man. When they mature and develop sexual feelings, a conflict develops between these feelings and the moral beliefs that have been implanted in their minds. Because of this, confusion and inhibition result. Children are seldom taught about the involvements of sex. They are not impressed with the seriousness of bringing a child into the world. Many are not taught birth control, which insures that children will not be born before they are wanted and can be cared for.

Atheists often believe that sexual morals are highly personal in nature and should fall within the category of personal morals because they do not interfere with the freedom of others. People seem to have an undue concern over what is happening in the world sexually. There are still those who contend that sexual intercourse is wicked unless accompanied by a desire for offspring. Because of the taboo against polygamy, many exhibit an undue jealousy towards their mates. It could be that other forms of misconduct are more fatal to marriage than an occasional infidelity. Sex drives are not limited to those who are married. However, there are societies that restrict sex to marriage. While many people are unsuited for marriage, they still have sexual needs.

Frigidity in women and impotence in men can often be attributed to fear, hatred and a dislike of or lack of trust in their sexual partners. This prevents them from giving themselves completely or acting spontaneously. There are probably as many types of love as there are people to experience them. It has been said that two people fall in love when they feel they have found the best partner available to them in light of the limitations of their own exchange values. The measure of love is the measure of one's assets for being loved. If love is present, it may add to sexual satisfaction, but sex and love are not inseparable entities.

Women, in general, have a greater sensitivity to emotions, such as love, than men. They are less capable of separating love and sex than men. Many women convince themselves that sex without love is impossible. As a consequence, they suffer heartaches when their sexual partners leave them, or they endure long periods of abstinence. Sex is the best exercise that some people get. It may be a contributing factor to the reason why married people live longer than single ones.

Of course, sex is not limited to intercourse between a man and a woman. Many parts of the world still consider homosexuality and the many other unusual sexual practices as crimes. Here, again, it is the author's belief that any practice among consenting adults is of a strictly personal nature and should not be restricted by law. It is also the right of those persons not believing in these practices to be protected from exposure to them. For example, the practice of flashing (revealing one's nude body in public) is certainly grounds for being jailed.

The author's feeling is that those who patronize prostitutes are wasting their money. Prostitutes are usually in business only for the money they receive. They have little or no emotional involvement with their customers. Consequently, they give very little in return for the money they receive. However, prostitution, which has been referred to as the world's oldest profession, is not going to be stopped by laws. At the present time, law enforcement officers spend a good portion of taxpayers' money on trapping and prosecuting prostitutes when they would be better employed on more serious crimes. If prostitution were legalized and licensed, the fees derived from it could be used for improving sexual health conditions and for promoting other worthy causes.

Movements have been under way in many countries to stop the production of pornographic matter. Other countries have, for centuries, already made its possession a crime. The author must, again, classify the possession of pornography as one of the personal freedoms. Persons who object to it do not have to purchase it. Those who receive pornographic advertising through the mails in the USA can fill out a form at the post office to stop further deliveries. It may be that the world needs better guidelines to prevent the strictly moral from being offended. However, let's not abolish personal freedom in the process. It might be noted that one of the largest collections of pornographic material in the world is housed in the Vatican.

Much has been said about the hotly debated subject of abortion. The Romans believed that the embryo was part of the mother, and until birth, the decision to have or not have an abortion was up to her. Many early people believed that a child did not receive a separate soul until it began to breathe. Up until that point, abortion was legally permitted.

There is no doubt that an embryo is a living thing. So is a sperm a living thing. Many billions of potential babies are deprived of existence through masturbation. It is the belief of many atheists that the embryo has virtually no knowledge of its existence. In fact, there are plants that may exhibit more of an awareness than an embryo. Of course, even a single cell living in some goop in a swamp exhibits a desire for continuance.

During early stages of development, the embryo does not have a memory with knowledge of life, death or pain. It is merely matter that exists without a real knowledge of existence. Whether it ceases to exist as an embryo or develops into a human being is of small importance to the matter comprising the embryo. The growth of an embryo falls into the same category as the growth of grass on the lawn. It is only after the growth processes have formed a memory system that continuance as we know it takes on importance. For these reasons, it is believed that the decision for or against abortion should be determined by the woman who possesses the embryo and, in special cases, by the man who caused the conception. The author does not believe in abortion. It is a pity that people are so dumb that they cannot stop making babies that they do not want. But that is not the way it is.

At this point, we might consider the morality of keeping people alive whose brain cells have been destroyed and who are little more than vegetables. Also to be considered is the morality of keeping someone alive who has no chance of survival. Often, the expense involved in keeping such individuals breathing is exorbitant. To compel families or the state to maintain life under either of these conditions is immoral in the eyes of the author. What value has continuance without hope?

Another misappropriation of law enforcement is the prosecution of marijuana users. Whether or not it is more injurious to one's health than cigarette smoking is debatable. One way or the other, people are going to smoke it. If it were taxed and sold in liquor stores, at least the state would be able to derive some income from marijuana users before they meet their untimely death. It seems everywhere today there are restraints to regulate and protect one's life. Is it better to live a long, regulated life or a shorter one with more liberties?

Nationalism, or a so-called duty to the state above one's duty to the people of the world, possesses, in the author's mind, a great threat to world peace. Too often, people blindly follow their leaders who profess the superiority of their state and its destiny to control the world, or a portion thereof. Of course, each area has its own beliefs, way of life, and problems that demand the most attention.

In a world where everything is moving towards sameness, it is becoming increasingly important to preserve the people's natural heritage. The educated people of so-called underdeveloped countries are, in the author's opinion, wrongly ashamed of their lifestyle. What's wrong with living in grass huts? In many ways, these people have a way of life far superior to that of the so-called "civilized countries".

It seems ironic today to hear talk of religious wars. Most wars now are of an economic nature. War usually makes the economy worse, since it accomplishes nothing but destruction. Could it be that the basic need in people to fight for survival is what generates wars? Possibly, what the world needs is an area set aside where people with these needs can go and fight and kill each other.

Man is a social animal. He has a need to overcome separateness. He accomplishes this by union with the group, by becoming part of the herd. In order to belong to the group, he must conform to certain customs. He must mesh with and gain acceptance among the masses. This need to belong and overcome aloneness dictates his morals, not God.

Many societies have set hard and fast rules by which the members must live. These societies often find it nearly impossible to mesh with other more liberal elements in the world. In a sense, it is too bad that the world cannot be fenced off, permitting each society to live in its own way. With the jet age shrinking the world, this becomes increasingly difficult. The necessity is growing for societies to live with each other and respect each other's beliefs.

There are many people today who have lost their faith. These people often show anxiety and have no aim in life, except the one of moving ahead. For their forefathers, salvation was the supreme goal; all other things were subordinate. Today, daily life has separated itself from religious thoughts. Man's main concern is material comforts. Young people, in particular, find themselves looking for a meaning in life and a purpose to live. The author cannot say that there are answers. There are many meanings in life. Everyone must find his own. As for a purpose in life, the author can think of no better one than to exhibit the properties of the matter that composes man. This matter desires to move ahead in a straight line or take the path of least resistance. Accomplishment is the result of motion. Accomplishment takes many forms. If one can find the form of accomplishment that brings happiness, then that person has found a purpose to live.

13. IN CONCLUSION

Difference in beliefs is often hard to comprehend. This is especially true when one considers people of similar backgrounds. However, each individual has different basic needs. To satisfy these needs, one makes assumptions. The hypotheses and theories that each individual accepts as truths are highly influenced by basic needs. The believer in God needs to believe in God, while an atheist needs not to believe.

In the day of Christ, the average man had little knowledge of science. It was only natural that he would conceive of religion in terms of the world he knew. In contrast, present day man is influenced by the scientific world around him. It was believed that science would lead man to the final truth. On the contrary, as man learned more and more, the truth became less certain. It became evident that if there is such thing as truth, and science reveals it, then the screen of man's limited senses prevents him from seeing more than a fraction of the truth.

It is only recently that science has developed the sophisticated instruments for looking at the minute particles of matter. The electron microscope, which can magnify 200,000 times, was only developed in the last century. Now we can see the 50 to 100 trillion cells that are in the human body. We can also see the 3 billion bases that make up the DNA that is in the nucleus of every cell. And we are learning what genes, which are lengths of DNA bases, code for particular proteins. The process of life is extremely complicated, but it does not need a man-made god for it to work. Because so much of the scientific knowledge is recent, the average human brain is still programmed to believe in gods.

As stated earlier, a characteristic of all matter is to have inertia. That is, matter tends to do in the future what it is doing at present. Some outside force, such as that exerted by other nuclear building blocks, is required to change the direction of motion. Life desires continuance because of the inertia of matter contained in living things. Most matter appears not to be aware that it exists. A brain is required to establish awareness. If matter does have an awareness of its existence, man's mind cannot comprehend that awareness. The desire for survival is a compelling force among all living things. This desire is one of the manifestations of the law of inertia. It is only natural that man would manufacture a soul to transcend his body after death and thereby gain the continuance that this law strives to achieve.

As we have seen, one of the basic beliefs in science is that matter can neither be created nor destroyed. If this is so, then the changing matter in our bodies cannot be created nor destroyed. Even though we cease to function as a whole animal at death, the matter composing our bodies continues on forever. In a sense, we are dying every moment because the matter in our bodies is dying or changing form from moment to moment. The dead matter is carried away out of our bodies. The dead cells are replaced by their facsimiles. In this way, the body continues to live.

Since gas, light and heat can be produced from solid matter, it can be said that they are also composed of matter. The human body can sense them. However, there are other forms of matter that the body cannot sense. Scientists have developed sensitive instruments that can record them. It could very well be that there are many more forms of matter in this universe of which man has no knowledge.

A general religious belief is that the human body is composed of matter that, in itself, does not possess the quality of life. Life is achieved through the existence of a soul that resides within a body. The soul is something without matter that remains with the body until death and then goes to Heaven or Hell to be with the souls of others who have also died. Some believe that other animals and plants have souls.

The author wants to point out that a major variance exits between religious and atheistic thought. To many atheists, the soul is nothing more than the function of the whole. There is no nonmaterial guide for the body. It works because the matter of the body exhibits life. In a sense, all matter is alive. The body is controlled by a brain with a memory and an extremely complicated network of communicating parts. Plants and animals are structured so that they are able to experience life and have feelings. Again, according to atheistic belief, there is no soul separate from

126

matter. The soul of anything animate or inanimate is the functioning of the whole.

People are inclined to think of a plant as lacking much in the way of sensory organs. Quite obviously, the world to a plant is different from that of a man. A plant may scarcely be aware of its existence. However, it has been theorized that plants are far more aware of what is happening than most people realize. The author has heard of experiments that were conducted by attaching electronic sensors to plants. When leaves were pulled from a plant, it sent out distress signals. Any number of people walked by the plant, and no signals were received. When the man who removed the leaves walked by, the plant sensed his presence and again sent out distress signals. Reality to a plant may be far different from man's reality, but at the same time, neither reality may be real at all.

As mentioned earlier, the birth and death of the human body is a continuous function. Because man has a memory, we might think of life and death as the beginning and the end of that memory. We can remember our childhood, yet as an adult, little or none of the matter that comprised our body as a child is still with us. Our memory is constantly altering its cell structure, and other organs of the body frequently replace their cells. Because the new cells are not reproduced exactly the same as the previous ones, the body goes through a change called aging. In other words, we, as a child, are dead. From moment to moment, at least part of us is being born and part of us is dying. The body continues to rebuild itself as long as it is able. At the same time, the rebuilding processes change, and various body parts may become destroyed. Finally, there are an insufficient number of working parts left to sustain the memory function. At that point, we say death has occurred. However, hair continues to grow, and other body functions continue, as long as they have nourishment.

Is the matter that comprises the body now dead? The author believes not. To him, it continues its existence in another form. Is a rock dead matter? If a rock is looked at under an electron microscope, one will see that it is alive with nuclear activity. The nuclear building blocks are far from dead. Is not the difference between animate and inanimate objects the degree and type of nuclear activity? Animal and plant matter are held by many to differ from other matter in that they were created by a divine power. However, it is possible to produce organic matter from inorganic substances in the laboratory.

Reality, if there is such a thing, may be very different from what humans experience. If there is something else that is real, and that something else is different from what man is able to experience, he has no knowledge of it. The author's belief is that reality is in the eye, that is, sensory organs of the observer. If we were constructed differently, we would perceive matter differently, and reality to us would be something else. The author contends that man today perceives matter differently than he did 2,000 years ago, and man today perceives matter much differently than did man one million years ago.

Recalling the case of the LSD users, they told us that their world became more colorful. They could see waves of heat and light. Somehow, an alteration of the brain had occurred allowing the individual to see things differently from what one would normally see. Again, is this a real or unreal world that they see? Isn't it only a different interpretation of matter, brought about by an alteration in brain structure that is no more real or unreal than man's normal perception? Since all people are basically constructed the same, they perceive the universe in much the same manner. Consequently, man assumes that this perception is reality.

Imagine a man composed of many billions of electrons, protons and other nuclear building blocks held together, magnetically, in some sort of mass. This mass has a memory, and sensory perception is accomplished by receiving electronic impulses. This man has no eyes, ears or other sensory organs as we know them. He might sense a tree as a branching mass of electrons, etc. radiating electronic impulses. A mountain would appear as a conical mass of the same. Man as an electron mass would have no knowledge of how man, as we now know him, senses electrons and other nuclear building blocks and nothing else. In other words, reality in this universe is how the observer senses it. There may very well be nothing that is real and true.

Everything experienced may be an interpretation of something that can only be called matter.

Some believe that nothing man experiences actually exists. They believe that the universe around us is a fiction of our minds. If such were the case, man would have a mind composed of nothing, conceiving a universe composed of nothing.

No matter what exists, one thing seems sure: as long as man must act in accord with the sensory inputs he receives, he is not free. He exhibits the capacities and limitations of his composition, just as a rock must exhibit the limitations that it possesses. However, man can think, and man can reason. It is through these thought processes that man often develops an unhappiness with his lot. The law of inertia, as in all matter, causes him to want to continue life. This cannot be. One day he will disappear, possibly without leaving a trace of his existence. The rock, as it appears to man, has the advantage of not being aware of its state. Man looks out at the universe. It appears to him to go on forever without meaning or justification.

Many religions of the world proclaim that man has free will. They state that it is because God respected man that he created man with a free will. The decision to live a righteous or a sinful life is left entirely up to man. However, when the brain is studied in detail, it becomes doubtful whether man has any true freedom in his actions. Chapter Ten has detailed the basic brain functions. Here, some of the other manifestations of thought are discussed.

The operation of the brain is dependent upon a steady stream of information coming to it from the outside world. Even with the vast amount of information that is stored in its memory, the brain is unable to function without receiving this outside information. In fact, mental activity becomes difficult or impossible when sensory inputs are reduced. The mind functions by comparing incoming sensory information to previously received and stored information.

Again, the functioning of the animal brain is dependent upon reception of information from the environment. The brain functions only in reference to information received from the body and sensory organs. There are no decisions made by the brain that are independent of sensory inputs. The actions of animals are controlled by these sensory inputs. It is only because of the staggering amount of simultaneous activity that occurs in the brain that man obtains the impression that he has free will. In actuality, every action is controlled by inputs reacting with established memory processes.

Body actions are related to electrical and chemical activity within the cells. Even though the complexity of interactions between body, nerve and brain cells is staggering, no evidence exists to show a spiritual or supernatural control over them. The body, nerve and brain cells react only to the laws governing electricity and chemistry. Again, nerve fibers do not carry sensations. They carry only patterns of electrical charges, which are deciphered by the brain. Brain tissue, in itself, does not have feeling. It can be operated upon without anesthesia. Fear, hate, love and other emotions are the result of membrane depolarization of appropriate clusters of brain cells. With the absence of the brain, no emotions are felt.

In attempting to relate reality to what man experiences, one must consider the apparent fact that all sensory inputs are distorted during the process of mental assimilation. The kind and amount of distortion is related to the brain's structure and memory processes. The memory does not store single items; it stores an interrelated collection of events. When these events are recalled, the cerebral processes do so in an orderly manner.

Now we are talking in general terms about how the brain functions. Nothing specific.

We previously talked about experiments that have been conducted on the brain. In some cases, electrodes have been implanted in the brains of both humans and other animals. By applying small electrical charges to the brain through these electrodes, body movements, speech and other reactions occurred. The individuals involved had absolutely no control over their reactions.

Some examples of reactions caused by electrically stimulating the brain are:

1. Stimulation of the hypothalamus caused one eye pupil to constrict while the other remained normal (monkey).

2. Stimulation of the right side motor cortex caused the left hind leg to be flexed. The action produced was in proportion to the electrical charge applied (cat).

3. Stimulation of the temporal lobe caused opening of mouth and movement of arm (monkey).

4. Stimulation of the red nucleus produced a turning of the head, walking on two feet, turning around, etc. Each time the experiment was repeated, a nearly identical reaction occurred. A stimulation of the red nucleus only 3 millimeters away produced only yawning (monkey).

5. Stimulation of the left parietal cortex caused the right fist to close. The contraction started with the first two fingers. The fist remained closed during stimulation (man). The patient commented that his left arm felt weak and dizzy.

6. Stimulation of the rostral part of the internal capsule caused a slow head turning and body displacement to either side as if the patient were looking for something (man). The patient considered the stimulated action as being spontaneous and each time offered a reasonable explanation for it. It was not determined whether the stimulation caused the movement that the patient was forced to justify or if a hallucination had resulted in the patient's need to look around and explore the surroundings.

7. A bull in full charge can be stopped abruptly in its tracks by radioing signals to stimulate the appropriate part of its brain.

These are a few of the results outlined by Dr. Jose M. R. Delgado in his book, Physical Control of the Mind.

These experiments are clear evidence to many that man does not have free will. All actions and interactions that animals experience are determined by sensory inputs reacting with the memory processes. It is due to the staggering quantity of reactions in process and the infinite number of ultimate effects that can be produced by these reactions that man receives the illusion of free will.

Because it is impossible to remove the mind from the brain and examine it, and a nonfunctioning brain does not appear to have a mind, the term "mind" becomes a term only to describe the functioning of the brain. It also might be considered that since it is impossible to remove the soul from a body and examine it, and a nonfunctioning body does not appear to possess a soul, then the term "soul" becomes only a term to describe the functioning of the body. If there is a soul, and mental activities are manifestations of that soul, then when the mind is manipulated by electrodes in the brain, the soul is also manipulated.

There are many who claim that they have seen and talked with God. The atheist is inclined to group these experiences in the category of hallucinations. Not a great deal is known about the occurrence of hallucinations. However, they appear to be unusual perceptions in the absence of peripheral sensory stimulations. They may be due to the recollection of stored information or an unusual interpretation of information entering through sensory inputs. Their activity appears to be associated with the temporal lobe area of the brain. It might be noted that electrical stimulation can produce hallucinating effects, such as conversations with God.

Present day man is attempting to direct a massive scientifically oriented and controlled world with a disproportionately small brain. The realization of the theory of relativity in physics is the consequence of the physiological structuring of man's brain. If that brain were structured in a different manner, a different theory might be realized. The human brain is not capable of realizing an absolute truth, if there is one. Any so-called truth or rule that the brain realizes is established in relation to a set of reference points gained through sensory perception.

The believer contends that the universe is regulated, at least to some degree, by a god or gods. The atheist is inclined to believe that the happenings are caused by interactions between matter.

Scientific reason is not perfect, but isn't it the best by which man has to live? If there is a meaning to life, that meaning is not clear, but why must life have a meaning? Is not the fact that life does exist enough? If there is a reason for continuing to live, it is to achieve happiness within the limitations of man's existence. Is there really any other choice? Man can condemn himself to despair if he takes himself and the world too seriously. There is little happiness in doing nothing, in thinking nothing and being nothing. It is the experience of life that generates happiness and a reason to live. It is an inherent property of all matter to move ahead and change through interaction.

Seventeenth century thinking seems naive today. Most likely in another few hundred years, twenty-first century thought will seem much more naive. In most cases, thinking of the religious world is still in terms of seventeenth century religious symbolism. Through prayer, man's courage and confidence in himself are increased. He believes that God is his helpmate. Science may supply man's livelihood, but it does not supply the emotional stability that man needs. As a result, he often accepts religion on emotional grounds, without evidence.

God becomes big brother. God alleviates man's fears, gives man emotional stability and promises continuance in Heaven. Because religion represents something beyond human and scientific experience, it needs no proof; it needs only faith. However, faith has become a hope, not a possession. Before the Christian era, man was quite willing to give up his life on Earth and join his ancestral gods. Now, man clings to life as long as he can. He is sure of life on Earth. He is not so sure of a life hereafter.

It might be interesting to note that some years ago, pollster Harris determined that of the 97% of Americans who said they believed in God, only 27% considered themselves deeply religious. The comment made by the majority was, "religion is good for the kids."

The modern concept of God is a regression to an idol image. The helping father or mother image of the middle ages has faded. Salvation is no longer the supreme concern to which other things are subordinated. Religion and daily life have, in many ways, separated. As fears became subdued by modern knowledge, the need for a God diminished. Whether or not God exists changes little in man's basic condition on Earth. The atheist believes that God is nothing more than the projection of man's unrealizable ambition for continuation. Many atheists do not completely rule out the existence of a god. However, they feel that if a god does exist, the form and existence of God is far different from that conceived at the time of Christ or that conceived at a time before Christ.

Science is not infallible; it does not have all of the answers, by any means. On the other hand, science has developed principles that appear to have validity, at least for the world in which man lives. It is through these principles that man's present standard of living has been achieved. Whether or not these principles are right or wrong, atheists believe they are the best that man has to live by.

It has been shown that the mind requires a continuous flow of sensory inputs to function. It cannot function other than in relation to these material impulses created by electrical and chemical activity. A spirit, on the other hand, has only an imaginary composition, which does not consist of matter. All evidence points to the conclusion that the sensory perception of animals is limited to material perception. That is, man cannot perceive God or other spirits if they do exist. What man does perceive is the concept of God that has been taught to him. If the spirit world cannot be known, then it becomes possible that there is no sprit world, and there is no spiritual existence beyond death.

Again, by definition, God is not composed of matter. Since force cannot exist without matter, God cannot exhibit force and have any control over man or the universe. In saying this, it is necessary to presuppose that God is not completely foreign to any concept that man might have of Him.

In recent times, primitive religions have been discovered among tribes of such places as New Guinea and the Amazon. These religions have survived for centuries, virtually unchanged. There are also written accounts of early religious practices dating back hundreds and thousands of years. By studying these primitive religions, it is easy to gain an idea of how religious belief began. In almost every instance, it has been found that gods developed from dead ancestors. If it is true that gods are nothing more than dead ancestors, then the probability is that there are no gods at all.

The authors of the Bible must not have been great intellectuals, considering the many contradictions written. Of course, the knowledge of their day was far from what it is now. At present, it seems incredible that a man, such as Archbishop Ussher, could be so sure of his chronology. Of course, he calculated it from the Genesis narrative, which was thought to be the word of God. According to Ussher, the creation of the world took place in 4004 B. C. at 9:00 a.m. on the 26th of October.

> *Ask and it shall be given to you; seek and ye shall find; knock and it shall be opened to you. For everyone that asketh, receiveth; and he that seeketh, findeth; and to him that knocketh, it shall be opened.* (Matthew VII:7-8)

This is a very nice reading. It lulls the mind into peacefulness. Unfortunately, in real life, it usually doesn't happen that way; that is, unless one is a king. Religion is taught. Man is not born with the idea of God. If it were not that man needs guidance, a reason why things happen, and continuance after death, God, religion and the church might have gone the same way as the idea that the world is flat. But I am told that as surprising as it is there are some people that still believe that the world is flat and has four corners.

Most religious people believe that there is a soul separate from the material body that transcends the body after death. Atheistic belief is that the body functions only with the matter it contains. There is no separate soul. There is no life after death. The purpose of life is to achieve happiness through accomplishment here on Earth.

14. BIBLIOGRAPHY

Adler, Irving. *How Life Began*. New York: John Day Company, 1957

Allegre, Claude and Schneider, Stephen, "The Evolution of the Earth", *Scientific American* (October 1994): 66-74

Allen, Grant. *The Evolution of the Idea of God*. London: Watts, 1931

Barnett, Lincoln. *The Universe and Dr. Einstein*. New York: Mentor Book, The New American Library, 1952

Contradictions of the Bible. Truth Seeker Company

Blatchford, Robert. *God and My Neighbor*. Chicago: Charles H. Kerr, 1934

Buchner, Ludwig. *Force and Matter*. Truth Seeker Company, 1950

Collins, Francis and Jegalian, Karin, "Deciphering the Code of Life", *Scientific American* (Decebmer 19, 1999): 86-91

Crick, Francis and Koch, Christof, "The Problem of Consciousness", *Scientific American* (Sepember 1992): 152-159

Damasio, Antonio R., "How the Brain Creates the Mind", *Scientific American* (December 19, 1999): 112-117

Damasio, Antonio R. and Damasio, Hanna, "Brain and Language", *Scientific American* (Sepember 1992): 88-95

Dawkins, Richard. *The Greatest Show on Earth*. New York, NY: Simon & Schuster, 2009

Delgado, Jose M.R. *Physical control of the Mind - Toward a Psychocivilized Society*. New York: Harper & Row, 1969

Doane, T.W. *Bible Myths and their Parallels in Other Religions*. New York: Truth Seeker Company, 1948

Fields, R. Douglas, "The Other Half of the Brain", *Scientific American* (April 2004)

Fischbach, Gerald D., "Mind and Brain", *Scientific American* (Sepember 1992): 48-57

Foote, G.W. and Ball, W.P. *The Bible Handbook: for Freethinkers and Inquiring Christians.* London: Pioneer Press, 1961

Frazer, Sir James. *The Golden Bough: A Study in Magic and Religion.* London: Macmillan, 1955

Freeland, Stephen J. and Hurst, Laurence D., "Evolution Encoded", *Scientific American* (April 2004)

Gershon, Elliot S. and Rieder, Ronald O., "Major Disorders of Mind and Brain", *Scientific American* (Sepember 1992): 126-133

Golden, Frederic and Lemonick, Michael, "The Race is Over", *Time Magazine* (3 July 2000): 18-23

Goldman-Rakic, Patricia S., "Working Memory and the Mind", *Scientific American* (Sepember 1992): 110-117

Gould, Stephen Jay, "The Evolution of Life on Earth", *Scientific American* (October 1994): 85-91

Hinton, Geoffrey E., "How Neural Networks Learn from Experience", *Scientific American* (Sepember 1992): 144-151

Ingersoll, Robert G. *A Few Reasons for Doubting the Inspiration of the Bible.* Austin, TX: American Atheist Press, 1997

"Is God Dead?" *Time*, vol. 87, no. 14 (8 April 1966): 40-45

Kandel, Eric R. and Hawkins, Robert D., "The Biological Basis of Learning and Individuality", *Scientific American* (Sepember 1992): 78-86

Keracher, John. *How the Gods Were Made (A Study in Historical Materialism).* Chicago: Charles H. Kerr, 1929

Kimura, Doreen, "Sex Differences in the Brain", *Scientific American* (Sepember 1992): 118-125

Koobs, Andrew. *The Root of Thought*. Upper Saddle River, NJ: Pearson Education, 2009

Lemonick, Michael D., "The Genome is Mapped, Now What", *Time Magazine* (3 July 2000): 24-29

Lemonick, Michael D., "How DNA Works", *Time Magazine* (17 February 2003): 50-51

Lepp, Ignace. *Atheism in Our Time*. New York: Macmillan, 1963

Meslier, Jean. *Superstition in all Ages*. New York: Truth Seeker Company, 1950

Oatley, Keith. *Brain Mechanisms and Mind*. London: Dutton, 1972

Orgel, Leslie E., "The Origin of Life on Earth", *Scientific American* (October 1994): 77-83

Russell, Bertrand. *The ABC of Relativity*. London: George Allen & Unwin, 1958

Russell, Bertrand. *Why I Am Not a Christian*. New York: Simon & Schuster, 1957

Sargant, William. *Battle for the Mind: A Physiology of Conversion and Brainwashing*. Cambridge, MA: Malor Books, 1997

Selkoe, Dennis J., "Aging Brain, Aging Mind", *Scientific American* (Sepember 1992): 134-142

Shatz, Carla J., "The Developing Brain", *Scientific American* (Sepember 1992): 60-67

Smith, John Maynard. *The Theory of Evolution*. London: Penguin Books, 1958

Stille, Darlene. *Animal Cells*. Minneapolis, MN: Compass Point Books, 2006

Stille, Darlene. *DNA*. Minneapolis, MN: Compass Point Books, 2006

Stille, Darlene. *Genetics*. Minneapolis, MN: Compass Point Books, 2006

Stille, Darlene. *Plant Cells*. Minneapolis, MN: Compass Point Books, 2006

Young, J.Z. *Doubt and Certainty in Science*. New York: Oxford University Press, 1953

Zeki, Semir, "The Visual Image in Mind and Brain", *Scientific American* (Sepember 1992): 68-76

THE DAM (Chapter 2)

Life, vol. 55 (25 October 1963): 30-41

Newsweek, vol. 62 (21 October 1963): 62

Saturday Evening Post, vol. 236 (30, November 1963): 76-81

Time, vol. 82 (18 October 1963): 43

U.S. News, vol. 55 (21, October 1963): 6

NOTES

NOTES

PAGE #'S AND DESCRIPTIONS OF ITEMS YOU'RE MOST INTERESTED IN

NOTES

NOTES

NOTES

NOTES

NOTES